D0538368

From the Redwood Forest

From the Redwood Forest

Ancient Trees and the Bottom Line: A Headwaters Journey

by JOAN DUNNING

with photographs by
DOUG THRON

CHELSEA GREEN PUBLISHING COMPANY
WHITE RIVER JUNCTION, VERMONT

For Chuck, whose love for the forest has guided this book. — J. D.

For my son Tristen, my favorite Headwaters hiking buddy,
for my parents Thomas and Lynda Thron,
and in loving memory of all the forests and animals of the Earth
that have lost their lives due to our thoughtlessness. — D. T.

Support for this book was generously provided by the Macon and Regina Cowles Foundation and Real Goods Trading Company.

Book design by Christopher Kuntze.

First printing August, 1998
00 99 98 1 2 3 4 5

Library of Congress Catalog-in-Publication Data
Dunning, Joan, 1948–
From the redwood forest : the battle to save the ancient trees of the Headwaters / by Joan Dunning : with photographs by Doug Thron.
p. cm.
ISBN 1-890132-11-X (alk. paper)
1. Old growth forest—California—Headwaters Forest Wilderness.
2. Old growth forest conservation—California—Headwaters Forest Wilderness.
3. Logging—California—Headwaters Forest Wilderness.
4. Dunning, Joan, 1948– 5. Maxxam (Firm) 6. Headwaters Forest Wilderness (Calif.)
I. Thron, Doug. II. Title.
SD387.043D86 1998
333.75'16'09794—DC21

Chelsea Green Publishing Company
P.O. Box 428, White River Junction, VT 05001
(800) 639-4099 www.chelseagreen.com

Contents

Acknowledgments ix
Prologue xi
A Photographic Journey by Doug Thron *(following page 180)*

Chapter 1 *A Beginning* 1
Chapter 2 *Cool* 4
Chapter 3 *Doug* 9
Chapter 4 *Commitment* 12
Chapter 5 *Pacific Lumber* 15
Chapter 6 MAXXAM 20
Chapter 7 *Gods* 24
Chapter 8 *Community* 25
Chapter 9 *Circling Headwaters* 27
Chapter 10 *Fog-Larks* 31
Chapter 11 *Going to Headwaters* 32
Chapter 12 *Headwaters Grove* 36
Chapter 13 *Witness* 38
Chapter 14 *Loss* 41
Chapter 15 *My Own Background* 44
Chapter 16 *The Third Fern* 47
Chapter 17 *History of a Redwood* 48
Chapter 18 *Fungi* 52
Chapter 19 *Flying Squirrels* 55
Chapter 20 *Owl Creek Grove* 59
Chapter 21 *Chartres* 64
Chapter 22 *Salmon* 65
Chapter 23 *Targets* 67
Chapter 24 *The Mahan Plaque* 73

Chapter 25 *Founders' Grove* 76
Chapter 26 *A Memory* 81
Chapter 27 *The Fiddle* 82
Chapter 28 *Ellen* 85
Chapter 29 *Clear-cutting* 88
Chapter 30 *Fisher Gate* 89
Chapter 31 *Allen Creek Grove* 93
Chapter 32 *Drawing Owls* 102
Chapter 33 *Meetings* 106
Chapter 34 *Kristi* 109
Chapter 35 *An Accident* 116
Chapter 36 *Houston* 117
Chapter 37 *Return* 124
Chapter 38 *Public Relations* 125
Chapter 39 *Jean at the Piano* 127
Chapter 40 *The Accidental Activist* 129
Chapter 41 *Slash Burning* 131
Chapter 42 *Plummeting* 132
Chapter 43 *Smut* 133
Chapter 44 *Headwaters Rally, 1997* 136
Chapter 45 *Mike O'Neal* 141
Chapter 46 *The Dipper* 149
Chapter 47 *Lumber* 153
Chapter 48 *Alternatives* 154
Chapter 49 *Toxics* 156
Chapter 50 *A Walk on the Beach* 157
Chapter 51 *Perspective* 160
Chapter 52 *High Pressure* 162
Chapter 53 *All Species Grove* 163
Chapter 54 *Soul Mates* 168
Chapter 55 *Shaw Creek Grove* 172
Chapter 56 *Salamanders* 174
Chapter 57 *Relics* 176

Chapter 58 *Expertise* 180
Chapter 59 *Camouflage* 181
Chapter 60 *Elk Head Springs* 182
Chapter 61 *Deluxe* 184
Chapter 62 *Weeds* 186
Chapter 63 *Morning* 191
Chapter 64 *South Fork of the Elk* 193
Chapter 65 *An Appointed Task* 199
Chapter 66 *Maidens* 200
Chapter 67 *Salmon at Grizzly Creek* 203
Chapter 68 *Luck* 208
Chapter 69 *Pepper Spray* 211
Chapter 70 *Spring* 220
Chapter 71 *Children* 225
Chapter 72 *Für Elise* 226
Chapter 73 *Democracy* 229
Chapter 74 *Return to Owl Creek* 231
Chapter 75 *Results* 237
Chapter 76 *Metamorphosis* 239
Chapter 77 *Knots* 241
Chapter 78 *Mythic Reality* 245
Chapter 79 *Butterfly* 247
Chapter 80 *Take-Out* 259
Chapter 81 *Ending* 261

Epilogue 263
Resources and References 266

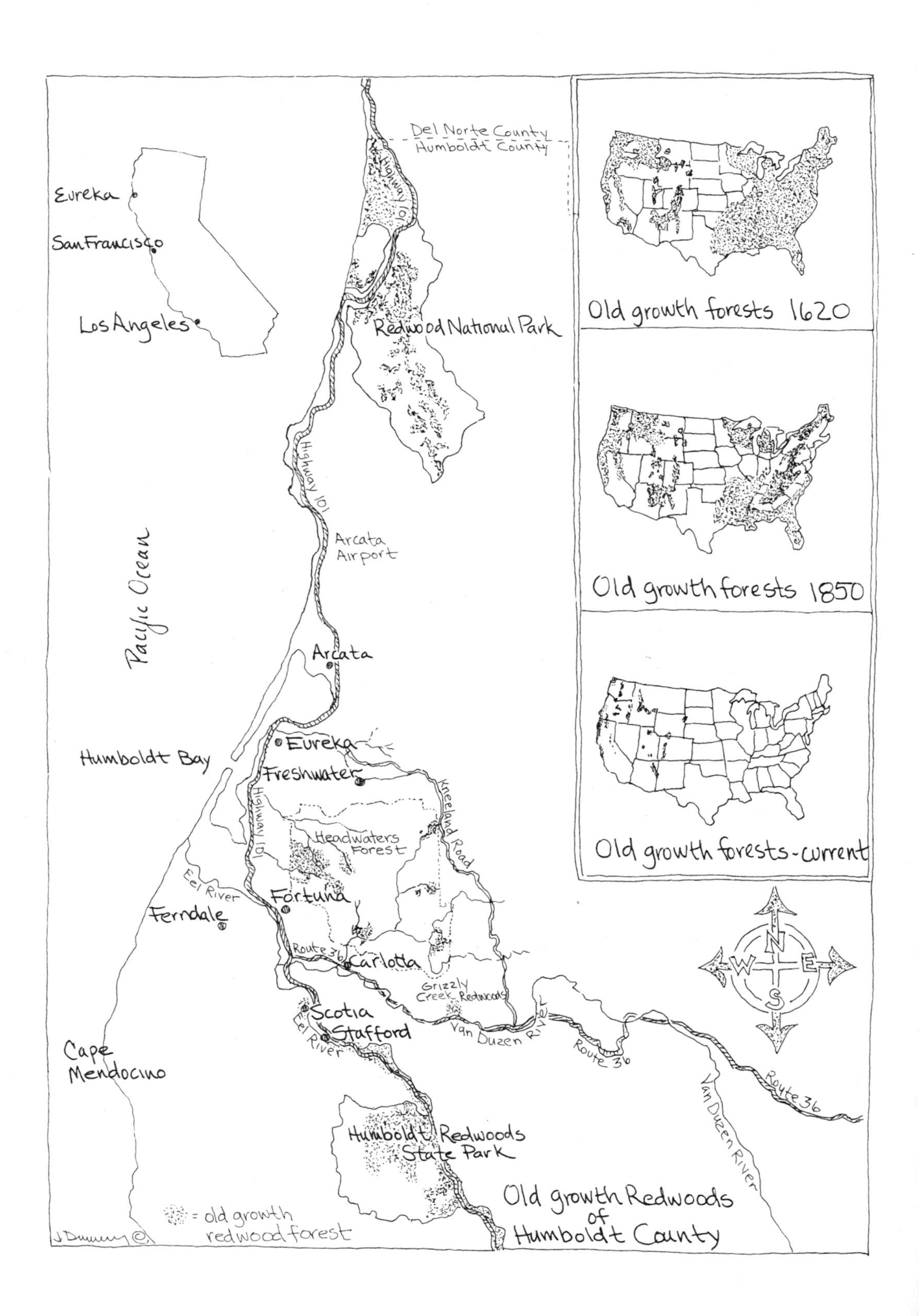
Del Norte County
Humboldt County
Highway 101
Eureka
San Francisco
Los Angeles
Redwood National Park
Old growth forests 1620
Old growth forests 1850
Old growth forests-current
Highway 101
Arcata Airport
Pacific Ocean
Arcata
Eureka
Humboldt Bay
Freshwater
Kneeland Road
Highway 101
Headwaters Forest
Eel River
Fortuna
Ferndale
Route 36
Carlotta
Grizzly Creek Redwoods
Scotia
Eel River
Stafford
Van Duzen River
Route 36
Route 36
Van Duzen River
N
E
S
W
Cape Mendocino
Humboldt Redwoods State Park
Old growth Redwoods of Humboldt County
= old growth redwood forest
J Dunning ©

ACKNOWLEDGMENTS

I would like to thank the dozens of people who have accompanied me on my hikes into Headwaters; Lighthawk and its pilots, for flying me over Headwaters well over a hundred times; Chuck Powell, for all his work helping to preserve Headwaters, and for countless hours spent working on this book; all the people who have worked so hard to preserve the entire sixty thousand acres of Headwaters; the Headwaters Forest itself for allowing me to experience a truly wild redwood wilderness; Lucy Ingrey, who traveled with me throughout the U.S. giving hundreds of lectures on Headwaters; and all the new people who will be inspired by this book to help save the endangered forests of the Earth.

Doug Thron

First of all, let me thank Charles Hurwitz. He has not chosen to host the public on his land, but nevertheless he has. Repeatedly I feel waves of gratitude and regret sweep through my mind, and I want to say thank you for giving me a reason to come to know the redwood forest in depth. In the battle to slow the unsustainable harvesting of the redwood forest, many people have found themselves, each other, and the forest. But I want to say, "Join us. Come know your forest."

As for those people who have given me monetary support, moral support, editorial support, information, advice—the forest gives its own thanks in the form of special dark and dripping moments in the company of ancient giants, and friendships deep with spirit and the mystery of the earth. I have been in service to the forest, and I feel confident that those who have helped me will already have been thanked by the forest directly. The forest seems to have an endless capacity to reach out and touch the lives of those who open themselves for a while to its service. I did not know this before I began this book. This has been the lesson.

Joan Dunning

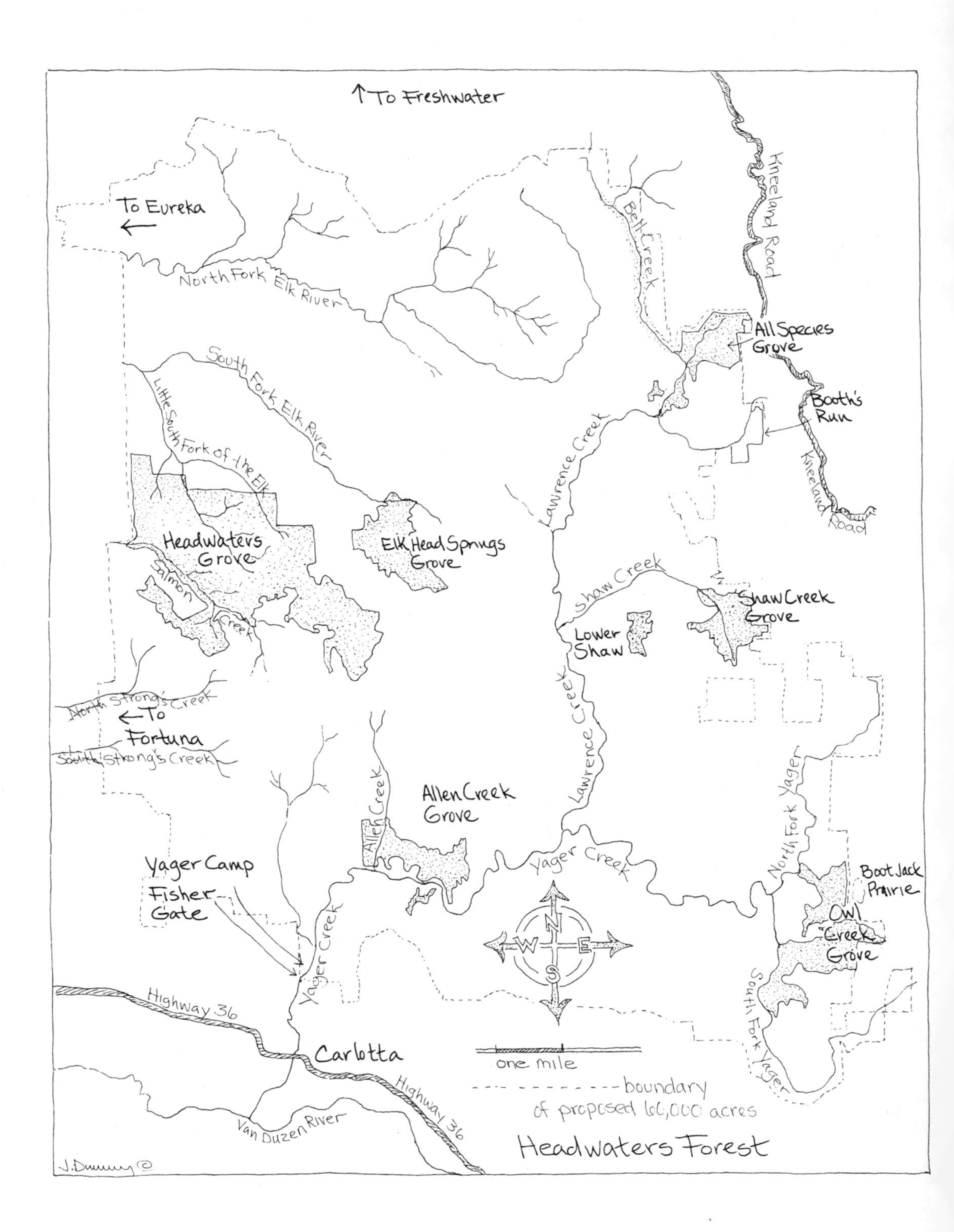

To Freshwater
To Eureka
North Fork Elk River
South Fork Elk River
Little South Fork of the Elk
Bell Creek
Kneeland Road
All Species Grove
Booth's Run
Lawrence Creek
Headwaters Grove
Elk Head Springs Grove
Salmon Creek
Shaw Creek
Shaw Creek Grove
Lower Shaw
North Strong's Creek
To Fortuna
South Strong's Creek
Allen Creek
Allen Creek Grove
Yager Creek
North Fork Yager
Boot Jack Prairie
Owl Creek Grove
Yager Camp
Fisher Gate
N
W
E
S
South Fork Yager
Highway 36
Carlotta
one mile
boundary of proposed 60,000 acres
Van Duzen River
Headwaters Forest
J. Dunning ©

PROLOGUE

Many people, when they think of California, imagine Hollywood or Malibu. However, coastal northern California, especially near the Oregon border, is as different from the usual stereotypical California as anywhere in the nation. It is overcast with rain or fog for much of the year. We receive an average of forty-five inches of precipitation annually. The natural vegetation is conifer, particularly redwoods and Douglas fir. In general, most people here do not swim in the ocean at any time of the year. The water is too cold, and the surf is too rough. The compensation is that we still have wildness here.

As for Headwaters Forest itself, most local people assume it is "far away" because it has received so much national publicity. I lived in Humboldt County for years before I ever asked, "Where *is* Headwaters, anyway?" I was shocked when the person turned eastward, surveyed the ridges behind Fortuna, and said, "See that tall stand of trees the third ridge back? That's the beginning of Headwaters."

I was surprised recently, when I took out a road map of Northern California, and studied the same area depicted on paper. There it was—the setting of my past year-and-a-half of thinking, writing and hikes. For some reason I had not expected to find Pacific Lumber Company's 210,000 acres there in the context of a common road map. I have viewed the land from the air countless times, explored it on foot, studied it on detailed geologic survey maps, or on company harvest maps that show all of the snaking logging roads, averaging 4.5 miles of road for every one square mile of "forest." Yet there it was depicted as simply a blank white space delicately crossed only by thin, blue lines designating Salmon Creek, Elk River, Freshwater Creek, Lawrence and Yager Creeks—the landscape known collectively as Headwaters Forest, within which lie the six old-growth groves that American citizens are doing their best to save—Owl Creek Grove, Allen Creek Grove, All Species Grove, Shaw Creek Grove, Elk Head Springs Grove, and Headwaters Grove. It is here that this book is set.

Douglas fir

CHAPTER I A BEGINNING

I NEVER intended to get involved with the controversy surrounding the Headwaters Forest, let alone write a book about it. I have always considered myself a naturalist, not an activist. For several years I have been working on a book about the birds of Baja. Yet this spring, instead of returning to Mexico to kayak around the islands in the Sea of Cortez observing bird life, drifting among dolphins, spending nights camped on beaches sparkling with bioluminescence at the waterline and stars in the sky, I have been tramping through clearcuts up to my ankles in mud. I have been flying over those same clearcuts in a small plane, circling again and again over landslides, slipping and sliding in silted creeks, walking along riverbeds choked with gravel, reading about the devastating effects of industrial forestry, and interviewing those involved with the destruction and salvation of Headwaters. Why have I done this? Sometimes I wonder what is wrong with me. Instead of helping me load up the car to head south through the deserts of Baja, my children have looked at my back as I have typed at my computer or joined me in the field for a crash course in forestry. They're not happy and neither am I. But this book is not about happiness. It is about yielding to conscience. It is about a forest, and it is also about us, about America.

I keep digging deeper and struggling higher to get a picture of just what I am writing about. I am struggling for wisdom and overview. I am trying to go from a novice on this issue on which people have been focused for over ten years, to an elevation from which I can describe the big picture. Each morning I lie in bed and I circle in my mind, a little higher, a little higher, like an eagle tapping into a thermal powered by the rising warmth of day. But, in fact, I lie in the predawn, when eagles are still grounded by cold. I rise through near darkness on a thermal of thought powered not by heat but by information. Each day I learn something new. Revelations lift me to a height I am not sure I even want to go, yet I scramble with questions to fill in any pockets and add to the force of what I perceive as truth.

I rise, surveying not just Headwaters; not just the buffer zone that surrounds it; not just the entire range of old-growth redwood rain forest that extends from the stunted forest of Arroyo Burro, south of Big Sur, to the huge, wet giants of Del Norte and Jedediah Smith; not just poor California, over-loved, over-admired, once endowed with riches; but America itself. I try not to let my gaze wrap around the globe to the darkness on the other side. America is enough of a topic. The sun has reached across the continent from east to west, fingering the Appalachians, sliding across the plains, halting and climbing the Rockies, fanning out across the deserts, scaling the Sierras, spilling across the ancient, inland sea of the Central Valley, and now it illuminates the street, beyond the blinds, beyond this little room where I crawled last night into my son's guest bed to tell him a story, and fell asleep midsentence. Without pausing, it reaches past us, down a few, short city blocks, out across the flats formed by ancient meanderings of the Mad River, over that last grasp of dunes, to be set free over the cool surface of the wide, lightening Pacific. America . . . the beautiful . . . what are we doing to you?

To the southeast, the sun has already slid down the damp slope of Boot Jack Prairie, and the male varied thrushes have slowed their early morning singing of mysterious songs that claim valuable old growth as their own, parceling it between them: "This . . . " is mine, "That . . . " is mine, "This other . . . " is mine. For thousands of years, song alone has been enough to claim whole trees, whole slopes of trees. The robin-sized birds have caught the sun on their orange breasts and sung ownership. "Mine . . . " and my mate's, ours . . . this canopy, the ground below, the dark spaces within the moss, between the twigs, beneath the leaves . . . and the beetles that inhabit them . . . the snails, the ants and wasps, the millipedes, the centipedes . . . shopping on foot down ancient aisles.

America the beautiful, God shed her grace, his grace, its grace . . . on thee . . . on thy thrushes, and on thy red tree voles and thy flying squirrels, and thy marbled murrelets and spotted owls, and on thy mycorrhizal fungi, extending the reach of roots to nutrients too microscopic to see. . . . I soar to get an overview, rising higher and higher, yet all the while that world beneath the thrushes' feet is opening deeper and deeper . . . the mysteries of the soil, the grace of soil, ten thousand years in the making . . . and how quickly sent downstream to clog the salmons' redds, forever lost. . . . I struggle to not vaporize in the vastness of mystery and its destruction.

This is a book about the satisfaction that comes from action, effective action, activism. I feel that America's "almost activists," the people who care but are afraid to act, are about to emerge. I know this only because I have reached a point in my own life when it is not enough to simply do "something," but when I have had to bring my actions up to speed with the destruction of my children's planet. I am never too far ahead or behind a mass of public realization. I write this book, yet before it reaches print, what it seeks to accomplish may have swelled up around me like a beautiful tide. If my heart has been pulled, I believe other hearts are being pulled by the same force. That force is conscience. The destruction of Headwaters is a personal issue. It is about the children who sleep in the bedrooms of our own houses. It is about the children we hold to our hearts and nurture to reach adulthood on a dying planet. It is about finally saying "Enough!" to people who accumulate more wealth than they can possibly count or comprehend in a sick corporate game of "bottom line."

This is the story of one person's journey behind the scenes of the destruction of one of the most magnificent ecosystems on Earth. The forces of that destruction and the counterforces that have held it at bay are so complex that the only way I can present what I have learned is by describing my own path of discovery and realization. I hope that in telling my story I will provide a model of action for the millions of parents who take care of every aspect of their children's lives but one: whether the Earth itself will survive. In my personal life, I reached the point of saying, I may be doing "something,"

I may be writing books about birds, about marbled murrelets even, but I am not doing *enough*. The forces of destruction are far outpacing my actions. If I take my children seriously, if raising them is not just some vanity for my own pleasure, then I have to act in proportion to the destructive forces at work on the planet. I issue this book as a challenge to parents (and I count us all as parents to our nation's children) to lift your heads, to assess and act effectively.

The young people whom I have met in the course of doing this book, people mostly in their twenties, beginning with Doug Thron, have been my guides. They stand, in my mind, with the majesty of old-growth redwoods. If I have a gut-feeling impression of the "terrain" of this book, it is of mountains put in a state of total disarray upslope, slumping to their knees, and, in the middle of this chaos, of young people emerging to stand straight and tall like the few stands of old growth that still remain. When they have so little reason to hope, they have grown clearer and fought harder, galvanized by war. I have been humbled and I have been heartened by the generation that is emerging between my children and me. They are my teachers. They are our hope. But the job is too big and too urgent to be left just to them, or we may as well cancel the piano lessons and the soccer camps because we are just fiddling while Rome burns.

CHAPTER 2 COOL

"WILL you just come to a meeting?" asked a friend who was in the process of founding an organization called Taxpayers for Headwaters after the September 15, 1996, Headwaters rally in Carlotta, California, which was attended by thousands of demonstrators from all walks of life. "We want to show Clinton that it is more than just Earth First!ers with dreadlocks who care about Headwaters."

At that point I was still determined to stay focused on my book about the birds of Baja, but my friend's request sounded simple, constructive, and limited. Mostly I said yes simply to get her to stop nagging me. I felt that it would be my token acknowledgement of responsibility to my local ecosystem. I went to that first meeting and the next thing I knew, I was up on a podium at a Headwaters rally on Arcata's central square preparing to read from a chapter on the marbled murrelet from my book *Secrets of the Nest*. People hear about marbled murrelets in the news, but few people actually know much about them. After I had briefly introduced myself and announced the formation of Taxpayers for Headwaters, a cheer went up. Cheering . . . I had forgotten about cheering. The sound woke me up and made me realize that I was actually acting on an issue about which I felt very deeply.

I heard my voice paint the scene of the controversy in my own words.

> *"Through a lush, old-growth rain forest in coastal California, a footpath makes a cool and winding cut through a delicate carpet of ferns and oxalis. Overhead, far above the world of humans, the canopy of this great forest is a private place, veiled in lichens, inhabited by flying squirrels and red tree voles. The early twilight of dawn is filtered wetly through the fog, and throughout the canopy there is constant dripping, as the needles comb the moisture from the air, gathering the tiny cloud droplets into larger drops that fall down through the trees, eventually watering the plants' roots."*

As I read, I heard my voice slow and dig in like a four-wheel-drive vehicle. All of the emotion I had been too focused to express, suddenly emerged and stood guard, seeing to it that my words were measured and hit their mark.

> *"Only a slight turn of its head reveals the murrelet's presence on the branch. Though the bird and the one glass-green egg it incubates are hundreds of feet in the air, there is no nest. Only a depression in the bed of thick lichens made by the weight of the small bird's body contains the egg."*

My love for this unusual, robin-sized sea bird, which should be a ground nester like the rest of its family but which, as a result of some bizarre twist in the course of evolution, wound up nesting high in the tallest living thing on Earth, forced the air out of my lungs with such compression that I felt like a tool at its disposal.

> *"How a marbled murrelet fledgling leaves its branch is still a matter of speculation. . . . To imagine plunging half-grown out of the top of one of the tallest living things on Earth, and flying twenty or thirty miles to an ocean I have never seen, untethers my imagination. It is the stuff of which perhaps not headlines but certainly myth is made."*

After I finished reading and stepped down, my friend found me and said, "The girl next to me said, 'This is cool.' " Cool. I was touched as I imagined her. I was glad I had been "cool."

Later in the rally, a member of Earth First! was introduced and took the microphone. He was clean-cut, hard eyed and serious. He told us that nearby, at that very moment, an ancient grove of redwoods called Owl Creek Grove was being harvested by Pacific Lumber Company. For days he had occupied a hammock suspended from the trunk of one of the giants, attempting to halt the destruction. Matter-of-factly he informed us that after the rally, he would sneak back into the forest, use his climbing gear to return to his perch, and resume the forest's protection. I stared at him as if he were a Druid, while he succinctly traced his path from the world of a nine-to-five job to his commitment to bodily defend virgin redwoods from destruction.

He described the fear of sitting high in a tree while men with chainsaws and bulldozers are working below.

"Most people are not aware of how the logging of these immense trees is carried out," he began.

I thought to myself, most people do not even *know* they're still being cut down. They think that all the virgin redwood forests are protected in parks. People have asked me incredulously, "You mean they are still logging those huge trees like you see in Redwood National Park?" Exactly. Only these trees, in Headwaters, are far beyond the roar of the highway, in some of the few watersheds that still spawn wild coho salmon.

"It is not just the cutting of the forest but the scraping of the forest floor by huge bulldozers to form the 'fall beds' that is so destructive. It is a unique characteristic of redwoods that, without a cushion of some sort to receive the weight of a falling old-growth tree, the trunk will shatter lengthwise on impact into tens if not hundreds of narrow pieces running much of the length of the tree. It used to be that loggers piled smaller trees and branches to cushion the fall. Now, with bulldozers, they excavate the forest floor itself."

I thought of the sorrel that carpets the floor of most redwood forests. I imagined a salamander one minute making his way over the forest duff, beneath his roof of sorrel leaves, and the next minute tumbling and suffocating in a rolling mass of mycorrhizal fungi and sorrel and ferns and rotting wood and dirt.

"Unless you have seen one falling, it is hard to truly appreciate just how big an ancient redwood actually is."

I thought of the Dyerville Giant, the tree that for many years held the title for being the largest redwood, until it suddenly fell during one particularly long and soaking series of winter storms. Few people knew it had fallen during the night. Without time to prepare, the rangers at Humboldt Redwoods State Park were not interested in coping with a traffic jam. Through connections with officials at the park, I found out the tree had fallen and went there, in the rain, with my children. Not only had it fallen, but because of the way that redwoods intertwine their roots in a sisterhood of trees, several other giants lay on their sides, their tops shattered by the force of the fall. As many times as I had taken people through Founders' Grove and stood while they looked up at this magnificent tree, I was not prepared for how big it would look on the ground. Wearing knee boots, we slipped and slid on what remained of the trail, looking into the muddy underside of the immense mat of roots. Roots that had supported this tree for two thousand years, since before the birth of Christ, were suddenly exposed, and we looked into the dark underside of time. Licorice ferns which had once grown on limbs twenty stories in the air were at our feet. Branches that had hosted generations of flying squirrels were now standing vertical, their squirrels launched in a flight that must be at the core of squirrel mythology. Perhaps we even made our way past limbs that had once held marbled murrelet chicks. In the winter rain, the overwhelming impression was one of immense size and age, as if a brontosaurus had slipped into the forest in the night and died, and we were the first to discover its carcass.

I thought I knew about the death of a redwood, until the speaker began describing the deliberate sawing down of an ancient giant. I was transfixed. My jaw was set, my eyes as riveted on his as his were on the people in the crowd. He described what it is like to sit night and day in a small hammock suspended from the trunk of his tree. I thought of climbers I have seen dangling for fun from the face of El Capitan in Yosemite. I am a cautious person. I hate heights. I hate doing illegal things. After years of farming, I one day realized that I hate being around large, powerful machines when they are operating. That alone would make me unsuitable for this fellow's self-appointed task.

"When the fall bed is made, they begin cutting. They have huge saws, *huge* saws," he emphasized. "They move up to the side of a tree . . ." I mentally held off what he was about to say, but he continued talking with the quiet steeliness of one who has hung alone, night and day, against the barked breast of a redwood under stars that beam just barely through the canopy. He glared at us, challenged us with his eyes, as he talked about the saw whining . . . and then suddenly he stopped, and described how, after it has been cut through and the saw is shut off, the tree still stands.

I savored his ability to communicate something so serious, at the same time that I felt like slipping from the crowd before the tree had a chance to begin teetering. I know that people commit large-scale acts of vandalism and terrorism every year. But the cold-blooded description of such an act by one who has witnessed it firsthand was too close for comfort.

"At first, nothing happens. The tree just stands there." I thought of this man hanging in his own tree, watching. I have carried his description with me every time I have gone to the redwoods and looked up along the trunk of an old-growth giant.

"And then it begins to just barely vibrate. It begins in the canopy and the trembling moves down the trunk. . . ." I felt my nerves trembling.

"It stands, as if it is still alive, but it is shaking. And then, slowly at first, it begins to lean."

CHAPTER 3 DOUG

A FEW days after the rally, I was listening to the radio while I was making dinner. The "Community Calendar" came on the air and the word "Headwaters" suddenly caught my attention. There was to be a slide show that very evening by a young man named Doug Thron. Perhaps I could lay the issue to rest in my mind if I went, either deem it important or dismiss it as mere political agitation by a few attention-getters. I spontaneously asked who in my family wanted to go. My twelve-year-old daughter, Suzanne, opted to stay home, but my nine-year-old son, David, and I quickly drank our soup and walked the few blocks to Humboldt State University.

We found the lecture hall, entered, and took seats. As we waited for the program to begin, I scanned the audience. Though I was born and raised in Southern California, I spent much of my adult life among conservative farmers and ranchers, first during fourteen years in Vermont, and then in the dairy community of Ferndale, where I moved with my children after returning to California. Only months before the rally and the slide show, I had put the finishing touch on my return by moving thirty miles north of Ferndale, to the university community of Arcata, in search of kindred souls with concern for the environment and love for the arts. But the fact was, while I was probably one of the most avant-garde members of my former communities, I was still conditioned by their conservatism. So it was with rather uneven feelings that I studied the audience around me.

Initially I was struck by how many of the young people had beards and dreadlocks. But then, in a change of heart, I began to wonder why *I* was one of the only people in the room over forty. Where were the "straight people"? Didn't this issue matter to them? Then I began to regard the audience rather maternally. Who were these children? Why so lost looking? What is the message with dreadlocks? Feral. This is the word they themselves use. Like goats or cats gone wild. I caught myself staring at one young man with red hair who was half curled in his seat, as close as one can get to fetal position in an upright chair that folds if one doesn't stay balanced. What was his background? How did he come to be sitting here, so deliberately un—un-what? Unlike his family? I felt a little unnerved by the crowd. For one thing, what was *I* doing there? If birds of a feather flock together, what did it say about me that I was surrounded by nose-ringed, dreadlocked, baggy-clothed young people? I could just picture my old neighbors standing by the podium, looking at *me* in the middle of this audience and shaking their heads.

I was bothered too, because there were so *many* of these young people, and I didn't see how any of them would be capable of "becoming President." The redhead in fetal position scratched himself under his arm slowly like a sleepy monkey, and I thought to myself, what are we doing? How can our culture be failing so many of our young people?

A professional fish biologist briskly stepped up to the podium to introduce the program. I was relieved. *He*, in fact, sounded as if he could go into politics. He was dynamic, articulate, and very upset. He talked about the once vast coho salmon population that has been almost completely destroyed in California as a result of poor logging practices, and how Headwaters forest is one of their last, precious watersheds with habitat still adequate for spawning.

Then he introduced Doug Thron, describing how Doug has been photographing the Headwaters forest for the past six years, trespassing on Pacific Lumber land for days at a time, often at great personal risk. I scanned the front rows for the most feral of them all, Druid with a camera. When Doug stood up, however, he was actually clean-cut, boyish. He wore a T-shirt, jeans, and well-worn hiking boots. His billed hat was pushed back on his head as if he had become accustomed to leaving it that way so he could look up at ancient trees. When he spoke, he was unselfconscious, highly focused, pausing momentarily from his work to inform us in the most economical language possible about what he had witnessed out in the forest. He was all business, and his urgency was infectious.

His photographs were shocking. Fading one slide into the next to the unearthly cadence of a Native American flute, he made whole forests turn to clearcuts and smolder and burn in slash fires. Aerial images of areas of pristine forest faded into images of the same areas, clearcut and bald, as if the hide of the Earth were infected with mange.

At times during the show I wanted to ask him to stop the slides. I had seen enough, I got the point, I was ready to act. But the show went on. I still try to fully imagine the experience of spending the first six years of one's adult life recording, single-handedly, at one's own expense, the actions of a multinational corporation as it hastily liquidates the last of the ancient redwoods behind the back of the nation and the world. The slide show was a record of Doug's life as well as the forest's death. After eons of silent, almost motionless existence, each branch, each trunk, even the floor of the forest itself, bound by the delicate growth of lichen and moss—this whole ecosystem was being destroyed because a Texas financier, Charles Hurwitz, head of a company called MAXXAM, had spotted assets in yet another corner of the world that had not yet been "maximized." Fueled by the quicksilver power of the junk bond, Hurwitz had swooped in, taken over the old Pacific Lumber Company, hired extra forces for a quick, terminal battle with the redwoods, and then, just as quickly, he would, probably, be gone, leaving the Pacific Lumber community to realize that it had been tricked into killing itself.

Doug has been described as a "war correspondent." When I recently read this description in an article I pulled from a pile of newspaper clippings about his work, I paused. Yes. Not only did this describe Doug perfectly, but it invoked, in two words, his task, the terrain of his work . . . the muddy, hard-packed logging roads; the clearcuts that go on for miles; the pampas grass choking out the young trees; the herbicide-sprayed scrub that stands suspended between life and decomposition, an eerie reddish brown over-parched earth; the skid roads; the erosion; the silted creeks bleeding with soil that was meant to be held in place by the Earth's skin of tree roots and moss and huckleberry and salal.

Without having seen the slide show, I might have imagined that Doug had given himself the enviable job of hiking about in silent, old-growth forest photographing tiny mushrooms and salamanders. But the fact is, he has spent six years witnessing the destruction of what he loves, documenting its passing. He reminds me of a person at a car wreck, holding the dying while they look out frantically to see if anyone will come. At the beginning of those six years, Doug says the forest was so vast that he could not predict for which tract MAXXAM's next Timber Harvest Plan would be filed. Then he went through a period of feeling superstitious. As soon as he began to wonder about a grove, it would be next. Now he does not wonder. So few groves remain, it is obvious.

Western trillium

CHAPTER 4 COMMITMENT

THE morning after I saw Doug's show, I decided to walk over to our local bagel shop and get some breakfast before I came home to work on my book on the birds of Baja. On the way there, a young man left a hairdresser's shop just as I was passing. It was Doug. His wife, Lucy, was having her beautiful blond hair blunted in a more practical cut, a "mom's do," as she called it. Doug was caring for their one-year-old son, Tristen, while the hair was lopped. We fell into stride, and I introduced myself. I had been thinking that it would be a good alliance if Doug did slide shows on behalf of Taxpayers for Headwaters, in order to increase membership. "Taxpayers" at that point was a fledgling organization of business and professional people who generally knew little about the Headwaters issue but wanted to learn more, and lend credibility to the cause. My idea was that Taxpayers could act as Doug's agent and pass the hat in order to pay him.

Doug and I sat out in the sun at the cafe and fell into a discussion of publishing. Doug said that he would like to do a book of his photographs of Headwaters. I still don't know exactly what got into me, but all of a sudden I heard my voice saying that I would be willing to write the text for his book if he would like me to. I also don't know why, once the words were out, I didn't quickly retract them. Doug looked at me and said that he would talk the offer over with Lucy. I figured maybe she would veto the idea, since he had told me that she was considering writing the text herself. But to my horror, they came to me the next day and solemnly told me that they would be very glad if I would write the text. I had given them my other two books to read. When I tried to disqualify myself, saying that I knew about the ecology of the redwoods, but not the politics, they countered by saying that they wanted "heart."

I agreed to write the text, consoling myself that I could research the subject quickly, do the writing in a few months, and get back to my other projects. It would be good practice in staying detached, "professional." I'm a desert person. I don't prefer the bones of the Earth fleshed out with vegetation. My father raised me to love the wind blowing unobstructed over the Mohave and to admire plants that survive there with only occasional precipitation. It would be easy to stay detached. I figured I could simply give an objective treatment of the ancient redwood rain forest; the impact of logging; more specifically, the philosophy of sustainable forestry as formerly practiced by Pacific Lumber Company; and the devastating effects of MAXXAM's arrival on the scene in what amounted to a hostile takeover. Doug said something about my visiting all six of the ancient groves within the mostly clearcut sixty thousand acres of the Headwaters Forest that environmentalists are trying to acquire. I nodded politely, but I thought to myself, "I'm not going to trespass. I'm not even going to get involved. A few trips to the library, a few interviews, and I'll have this book done."

As soon as I had agreed to the project, however, I was confronted by an unpleasant wall of stereotypes and conflicting propaganda. I found myself asking, "What *is* Headwaters?" Is it long-haired protesters photographed for the newspapers? Is it angry loggers fearing for their jobs? Is it billions of dollars pushed back and forth across the table between the government and a high-rolling financier? What about bizarre reports of attempts to trade Department of Motor Vehicles offices and San Francisco Bay's Treasure Island for redwoods as part of a confusing, ever-mutating agreement between the government and Hurwitz, called "the Deal"? Is Headwaters simply an expensive preserve for an obscure little sea bird called the marbled murrelet? Or is it an earthly wonderland right here on Earth, more mysterious than any fiction, barely understood, abused because it is out of sight?

Not only is Headwaters itself far from public roads, behind locked gates, but most of the significant players are reclusive. Salamanders are camouflaged to the damp duff of the forest. Marbled murrelets come and go at dawn and dusk. Red tree voles live for generations in forest canopy two hundred feet off the ground. Mycorrhizal fungi reach their root-extending filaments out of sight beneath the soil. Spotted owls stare silently, camouflaged against the trunks near which they perch. Loggers and their families watch events transpire without speaking out, while sharing private dread that at the current rate of logging, they too are candidates for endangered species status. The private citizens who have dedicated the past decade or more to protection of the redwoods spend much of their time behind the scenes commenting on Timber Harvest Plans, attending meetings, writing reports, researching lawsuits. Even Charles Hurwitz, the man whose company is clear-cutting the redwoods, is invisible. Not a local resident, not a woodsman, he directs the MAXXAM operation from Houston, granting few interviews.

Contrary to my plans, this book is a journey into that hidden world. In between trips to the library, I developed an increasingly personal relationship with this controversy known as Headwaters. As I began interviewing people and actually seeing from the air what is happening to the land, as I slogged through the mud of clearcuts I never intended to cross to reach groves I never intended to visit, the forest began to tell me its own story. This book is an example of how a project with an innocent beginning can envelop the soul. For each of us, regardless of where we live, there is a river, a mountain range, a beach, a whale, a peregrine, a gnatcatcher, that, if we merely give our time as a witness to loss, will gradually unite the veins of its existence with our own, will ground us by putting us in touch with what is wild and speechless, will empower us when we speak out in defense of the powerless.

CHAPTER 5 PACIFIC LUMBER

FOR a while, I thought of Headwaters as the elephant in the parable about the blind men and the elephant. In the story, an elephant is standing, surrounded by blind men who cannot agree on what an elephant really is. One holds the ear and says with certainty that an elephant is flat. One holds the tail and says with certainty that the elephant is like a snake. And one holds the leg and says that an elephant is like a great tree. I used to think that the breakdowns in communication between demonstrators and police and Pacific Lumber officials occurred because we were all "blind men."

Now I do not think that we are looking at the work of blind men. In the face of reality, the parable has lost its meaning like a child's story that has been outgrown. Now that I have seen the ravaging of the land that defies decency, that bears no resemblance to forestry, I view Headwaters as a hostage in a takeover by a man, not a forester but a speculator operating from Houston, who is stripping the surrounding land bare while he holds the remaining groves for ransom.

I think that many well-intentioned people work for MAXXAM, and that they have been fooled into believing that they are still working for the old Pacific Lumber Company, and that rural simplicity, decency, and pragmatism are still in effect. For the old Pacific Lumber Company to liquidate its own forests would have been to commit suicide. To treat its own workers without respect would have been counterproductive. Pacific Lumber, like the families on the ranches that surround the town of Scotia, was in for the long haul. Its owners and bosses worked their way up through the company in partnership with their community, and they were there on the spot to be accountable for their actions.

In a 1981 letter to stockholders, Pacific Lumber Company's board of directors wrote, "corporations have an obligation to society as a whole. In particular, companies charged with stewardship of scarce resources such as timber have a duty to use such resources wisely. . . ."

I have the impulse to pause here, to allow you time to reread those first two sentences and let their elegance stand alone in light of what is happening today.

The letter goes on, "Continuity and stability of management are necessary for the intelligent development and harvesting of the company's timber resources. . . . The Company was a pioneer . . . in developing and applying the continuous yield forestry principle . . . complemented by . . . selective harvesting of its timberlands to insure the continued productivity of these lands to perpetuity. . . ."

The authors of this letter then looked into their crystal ball and wrote, "Uninvited attempts . . . to take control of the Company and institute a management philosophy which would seek a dramatic increase in the short-term yield of the Company's redwood-producing properties would have a destructive long-term impact on the Company, the Company's stockholders, employees, customers, suppliers, and the communities where the Company operates."

In 1985, the predators of corporate America finally caught up with Pacific Lumber. A company cannot sell its stock on the open market and escape notice. In that year, Charles Hurwitz spotted a company whose huge assets were out of proportion to the relatively modest annual dividend it was paying its shareholders. Even though he himself did not possess the money to buy the company, and even though the purchase was dependent on the rapid and nearly total liquidation of a vast and ancient ecosystem, Hurwitz moved in for the kill. The vast tracts of forest that Pacific Lumber had once stewarded were about to be "maximized." Albert Stanwood Murphy, past CEO of Pacific Lumber, in the grave just thirteen years, would not have been fooled by corporate stealth. A. S., as he was known, took the reins in 1931, a decade after his grandfather, a Detroit millionaire, bought the company. The Depression was on, and most lumber executives were easily seduced by the quick dollar to be made as the internal combustion engine replaced steam power, revolutionizing logging. This revolution came in three forms:

The mobile Caterpillar tractor replaced the old, stationary steam donkey, which had required elaborate cable systems to move cut logs. In contrast, the Caterpillar could go almost anywhere. It made its own roads and then used them to pull out the logs. The practice of making fall beds from the branches of noncommercial trees, leaving the soil of the forest intact, was abandoned because the Caterpillar could excavate the earth itself, moving the mass into layout after layout to cushion the fall of successive trees. The Caterpillar's impact was devastating. Using this machine, more than three-quarters of the surface of the land at a logging site may be disturbed. Often up to one-half of the area is consumed by roads and skid trails.

At the same time, the ax and the cross-cut saw were replaced by the chain saw. Where it had taken two men two to three days to fell a twelve-foot-diameter tree, suddenly one man with a chain saw could single-handedly cut down numerous trees in a day.

Finally, the logging truck replaced railroad flatcars. The tedious job of laying down track to each new logging site was no longer necessary. With the Caterpillars making the roads, the logging trucks could then reach once inaccessible timber swiftly and with ease.

Needless to say, logging companies were dazzled by this new mechanization, but A. S. Murphy had the wisdom to step back and project the future

Coast redwoods

impact of such efficient technology. At a time when clear-cutting was the rule, he adopted policies which were to make PL a leader in forest stewardship. As a woodsman and a hunter who hiked his company's forests, he saw clearcuts as an indignity to the land, animals, and people of the area. California lawmakers, appalled by the sweeping changes brought by the internal combustion engine, instituted tax incentives for those lumber companies willing to harvest their lands with moderation. Instead of stripping PL's land, A. S. Murphy instituted "selective cutting" as a company rule. This policy limited harvest of a stand to 50 to 70 percent of the mature trees, opening up the canopy so that the younger trees put on rapid growth. After fifty years, these trees, which had served a dual function of holding the forest's soil in place and preventing the invasion of sun-loving shrubs, could then be harvested, leaving behind the next generation of young trees to in turn close the canopy.

Under A. S. Murphy's leadership, the company followed a policy referred to as "sustained yield." This means that the annual cut of the forest would not exceed a year's growth. While it may seem like common sense that one should not spend one's principal but only the interest in any given year, Pacific Lumber stood alone in its wisdom. Before the takeover in 1985, the pride and loyalty which employees of Pacific Lumber felt was almost palpable in the small company town of Scotia. David Harris, in his fascinating book *Last Stand*, about the MAXXAM takeover of the Pacific Lumber Company, describes one mill worker's affection for PL.

> *On the way to work from the coffee shop, he passed hundreds of men going home, smelling of sweat, fresh tree sap, and diesel fuel. Once such parades were a sight all over Humboldt County. Lumber companies and sawmills used to be as thick as fleas on a dog, but not anymore. What was there, five or six of them left in Humboldt now? There'd been more than a thousand once. There were still several dozen left between here and Canada, but none of them had lasted the way PL had. Because PL was never just any lumber company, at least not since the Murphys had owned it. It was, Kelly always pointed out,* the *lumber company. The rest, they'd just cut the shit out of the*

forest until they didn't have no more trees and then they'd move on someplace else. Not PL. As he liked to say, fucking PL was made to last forever. They cut slow and easy, so there'd always be trees left. By the time all the first growth was gone, the second growth would be just as big, and on and on like that forever and ever. The way the Murphys set this place up, there'd always be work. That alone made it one of a kind.

Another book, *Life in the Peace Zone*, a photographic portrait of Scotia created by Hugh Wilkerson and John van der Zee in 1971, conveys this loyalty in more innocent terms in a description of the Scotia school:

When Peggy Rice asks her class, "How many of you would like to live in Scotia and work for Pacific Lumber when you grow up?" nearly all the boys raise their hands. And so do half the girls.

Wilkerson's portrait goes on:

Though the town's population is around a thousand, it has been at that figure for most of this century, and for many years Scotia was second only to Eureka in size; yet apart from the owners of the company itself, Scotia has yet to produce its first millionaire, scientist, inventor, philosopher, composer, painter, author, or big-league ballplayer. Where likenesses are valued, there may be little competition, but there is also little pressure to excel. . . . What Scotia is really offering those dismayed with the world outside, the tie that pulls men back who vowed to leave, is not the promise of fulfillment but an assurance of moderation, the possibility of living a humane life in a humane community. And for that, there will always be a waiting list.

Before the 1985 takeover, while other Humboldt County timber companies had liquidated most of their ancient forest, Pacific Lumber still had lush old-growth forests that spread for miles beyond the town of Scotia. The groves that are now under dispute—Headwaters, Elk Head Springs, Shaw

Creek Grove, All Species Grove, Allen Creek Grove, and Owl Creek Grove—were only a small fraction of a vast sweep of pristine ancestral forest. Friends of mine who explored these forests before the MAXXAM takeover say they had to be careful not to get lost in the miles of old-growth. Now one hikes for miles through desolation to reach the remnant stands.

CHAPTER 6 MAXXAM

ACCUSTOMED to providing information to reporters and inquisitive audiences at a moment's notice, Doug can summarize the history and do the math of the MAXXAM takeover of Pacific Lumber as though they were second nature to him. While I believe him when he says that it has been hard for him to grasp the complexity of the issue, I still don't believe that he can possibly have struggled with the details the way I have. But finally the overwhelming convolutions of facts and figures are beginning to congeal into manageable concepts.

It was in 1985 that Pacific Lumber's exceptional assets attracted the attention of Charles Hurwitz, a financier and corporate raider on the prowl for undervalued companies worthy of purchase and "reorganization." In a surprise virtual takeover propelled by the sale of high-interest junk bonds, Hurwitz, under the protection of his Texas-based corporation, MAXXAM, seized Pacific Lumber before the people of Scotia knew what had hit them. Precisely because it had attempted to steward its land, protecting assets for the benefit of everyone from murrelets to loggers to stockholders, Pacific Lumber had become vulnerable.

Suddenly, trees that had lived for unnumbered eons were surrounded by people making calculations. In order to buy the company, Charles Hurwitz incurred approximately $864 million in high-interest debt. To pay the interest on this debt and still pay himself a handsome profit, he almost doubled the workforce, hiring outside loggers and accelerating the rate of cut of the forest to three times what the old Pacific Lumbe had deemed sustainable. In addition, he liquidated $60 million out of Pacific Lumber's generous pension fund and sold a large office building in downtown San Francisco.

Meanwhile, payment on the principal of the debt has not been made. To put the finances of the takeover in the simplest possible terms: over the past ten years Hurwitz has milked over a billion dollars out of Pacific Lumber Company assets, yet in the next six years he will have to raise another billion to pay his creditors.

Many judge that there is not this much standing timber left. In addition, watersheds are already strained and crumbling, communities downstream are outraged, and there is growing suspicion that Hurwitz will eventually simply sell the company, leaving Pacific Lumber Company, the largest employer in Humboldt County, in ruins. Unknown to many people is the fact that Hurwitz has divided Pacific Lumber into three parts: Pacific Lumber Company, Scotia Pacific (SCOPAC), and Salmon Creek Corporation. While the debt is consolidated in Pacific Lumber and SCOPAC, which are still owned by MAXXAM. Hurwitz himself owns the timber-rich Salmon Creek Corporation, which remains conveniently debt-free, and holds title to the old-growth groves of Elk Head Springs and Headwaters. Thus, it is not as though Hurwitz himself will be ruined. This is the beauty of a corporation. When "misfortune" strikes, one needn't change one's country club or even hang one's head, especially if one does not live locally to view the consequences. One can simply "blame it on the environmentalists."

One might ask at this point how a wise company like Pacific Lumber could allow a takeover to occur. Scotia lies behind what is called the Redwood Curtain. It is isolated by its own forests. That isolation produces a pride in its people that arises out of self-reliance. Until 1912, overland travel from San Francisco to Scotia, a distance of approximately 250 miles, was extremely slow. The company's principal contact with the outside world was through Humboldt Bay, thirty miles to the north. Thus Pacific Lumber remained a self-sustaining entity long after most companies had taken advantage of the conveniences of the twentieth century. The company treated its own water, generated its own electricity, manufactured its own replacement parts for its machines, and even built many of its own machines in

Scotia. But isolation not only can produce self-reliance and pride, it can result in arrogance or naïveté. Many believe that the board of directors of Pacific Lumber simply did not believe that a takeover of such a locally secure company was possible. In addition, A. S. Murphy's widow, Suzanne Beaver, and her sons, Woody and Warren, found that they were no match for the slick, big-city lawyers and financiers with their voluminous reports designed to defy understanding. Old-fashioned honesty, no matter how tight-grained, was as defenseless as the ancient trees themselves. While Pacific Lumber's ethics had been preserved by isolation from the twentieth century, Charles Hurwitz's tactics had been honed in a cutthroat and impersonal world where ethics are viewed as merely quaint.

Indeed, Charles Hurwitz has already cost U. S. taxpayers $1.3 billion due to the federal bailout of the failed United Savings Association of Texas. During the early years of Reagan's presidency, the savings and loan industry was deregulated so that investors' money was no longer backed by actual recoverable assets. Depositors' money was still insured by the Federal Deposit Insurance Corporation, however, making them vulnerable to what the *Wall Street Journal* described as a "spider web of deep pocket investors" looking for free money. In 1982, junk-bond king and soon-to-be-felon Michael Milken helped Hurwitz to gain control of United Savings Association of Texas, which in turn purchased junk bonds from Milken's brokerage firm, Drexel, Burnham and Lambert. When United Savings Association of Texas investments failed, it was the American people who repaid the "loss" to its investors.

Though the FDIC and the Office of Thrift Supervision had already filed lawsuits against Hurwitz by the spring of 1994 for his actions in the scandal, this did not prevent Hurwitz from being on the cover of the April / May / June, 1994, issue of *Leaders* magazine. Perhaps you have never heard of *Leaders*. I hadn't, until I was handed a copy by a friend. Charles Hurwitz's face beams from the glossy cover of this substantial and downright lavish magazine. I looked inside the cover page and read, "Circulation is strictly limited." Oh. It went on, "To receive *Leaders* magazine one must be the leader of a nation, an international company, a world religion. . . ." No wonder. I do not know any leaders of nations, international companies, or world religions. But one can also receive it if one is a leader of an "international labor organization, or a chief financial officer, a major investor on behalf of labor or corporate pension funds."

The copy for the cover photograph reads: "Charles E. Hurwitz, chairman, president and CEO of MAXXAM Inc., understands that it takes a long-term commitment to value to build success. MAXXAM's value has evolved from a thorough commitment to leave no stone unturned."

When I read this last sentence, all I can envision is the gravel that is choking the riverbeds of the Van Duzen and the Eel, washed down off slopes that have been stripped of their soil. I see the communities at Elk River, Freshwater, and Cummings Creek, overwhelmed by the mud and gravel that have been turned loose upslope in their ravaged watersheds. I think of the community of Stafford, buried beneath a massive mudslide after MAXXAM clear-cut the nearly vertical face of the mountain above. From the air, it is obvious that the stones that are not being left unturned, that are destroying roads and houses and gardens and peaceful lives, are simply a necessary casualty in Hurwitz's grand plan for harvesting the forests. And this is only the beginning. The watersheds have only just begun to unravel.

The paragraph ends with, "Beginning on page 48, this master builder of value and profits examines the process—which he is determined should be fun—of building success." Fun? I thought of the little tricycle that Doug photographed at the Stafford slide. The child's house had been buried along with much of the rest of the community. The tricycle had ridden on the top of the debris torrent that swept away some houses and filled others up to their ceilings with mud. Apparently the residents knew something was wrong when the creek abruptly stopped flowing. They evacuated in time to save their lives before the mountain, no longer held together by vegetation, turned loose and obliterated their community. Fun?

In the interview, Hurwitz is asked, "MAXXAM operates three core industries, and your Kaiser division is world famous, yet the MAXXAM name is probably the world's best kept secret. Do you plan to change that? How will you make MAXXAM a household word?"

To which Hurwitz replies, "It's not a high priority to make MAXXAM a household word."

In fact, it is integral to Hurwitz's operation that the people of Humboldt County continue to believe that the old Pacific Lumber Company is still alive and well. I believe that if the proud name PACIFIC LUMBER COMPANY, long emblazoned on the side of the old-growth mill at Scotia and visible from the highway, were painted out and replaced by the name MAXXAM, it would spark the beginning of a major change of awareness in this county, perhaps depriving Hurwitz of this final period when the remaining acres of ancient redwoods, once carefully meted out by Pacific Lumber, are hastily cut down so that Hurwitz can *pretend* he is going to pay off his debt.

Numerous private and governmental lawsuits have been filed against Hurwitz, MAXXAM, and Pacific Lumber. Thus, many people wonder why someone who already owes the taxpayers $1.3 billion continues to call the shots in the government's "Deal" to acquire the Headwaters Grove. Many people are perplexed at how the "debt-for-nature swap" that is intended to cancel out some of Hurwitz's debt to the citizens of the United States in exchange for ancient forest, has been turned around so that we are actually offering him *more* forests to cut.

CHAPTER 7 GODS

I HAVE just returned from a small grove of redwoods by the side of the road, a part of Grizzly Creek State Park, which I visited with a long-time resident of the area and his nine-year-old son. We went to see southern torrent salamanders. The old names for this salamander are the seep salamander and the Olympic salamander. We went to a seep. From beneath the soil at the top of a cliff, water spilled down a wet face of gray sandstone into a small pool. Five finger ferns bobbed and dripped as they were hit repeatedly by the splashing water. Beneath the little fall on the forest floor, the leaves were shades of wet brown.

"I don't like to disturb anything," the man said. "I just gently lift the rocks and leaves and put them back."

We delicately peeked through the leaves, squatting down near the splash zone, until he said, "Here's one."

He wet his hand in the fall and delicately picked up the salamander.

"I have never seen one so small," he said.

It was only about an inch and a half long. It strode across the wet finger, suddenly in the light, suddenly high in the air, and it reminded me of an infant and of a god. An infant, I guess, because of its rounded, nubby fingers and its big eyes and its look of trust. But, "Why a god?" I asked myself. The words "Because of its permeable skin," came to me as my reply. God of the underworld, I thought to myself. I was suddenly conscious of the ground beneath my feet. What beauty might I be standing on?

The man's son wet his own little hand and took the salamander from his father. His moves were easy. He was as at home by a seep as his father was, yet he had two gods—a salamander and a man.

Southern torrent salamander

CHAPTER 8 COMMUNITY

ONE night a man came and spoke to Taxpayers for Headwaters about the history of corporations. He said that originally corporations were chartered only for specific periods of time and specific purposes, and renewal of these charters was not guaranteed. In 1886, the Supreme Court breathed life into corporations, giving them corporate personhood. Unlike an unincorporated business, whose owners maintain personal accountability, a corporation exists abstractly, on paper, like a humanoid with no heart.

As a collective public, we no longer have effective authority over the actions of that robot. Individuals stand up periodically and object, government agencies lob fines like pebbles, but we have lost our power to say, "Sorry, you are not conducting yourself in a way that promotes the sustained health of the planet. We are not renewing your charter." As long as we continue to focus on "regulation," a system in which corporations have every financial and legal advantage, we will not be able to recover democratic control, and without control of our corporations, we have little control over the conditions of our environment and the fate of our communities.

I lay in bed this morning and tried to lift a little higher in my thoughts. We are all interconnected, whether our lifeblood flows through veins made of rivers, or xylem and phloem, or gas lines, or telecommunication lines, or currents in the ocean, or the currents that flow through the human mind and heart. I refuse to separate any segment of this system and excuse or condemn it. I simply want to understand. I rise on a thermal of questions. What is wrong with this system that a place like Headwaters is bleeding its soil, its infant trees?

I remembered the screech owl I had heard the night before at Humboldt Redwoods, filling the forest with its ghostlike calls. I began to wonder if Charles Hurwitz has ever camped out in the open, without a tent, in a redwood forest. Is this a ridiculous question? Is it rude or ridiculous to wonder if a person has ever slept out on the ground, alone, and listened to a screech owl in the night, coursing its way on silent wings through giant trees? Does one have the right to kill what one does not even know?

To strangle someone, one must feel the beauty of the muscles. One must feel the potential of the breath. In each second, as one puts that life to rest, one must make a choice, choosing until there is no turning back. In a drive-by shooting, one simply holds a gun and aims and does not know the before or after. Could Charles Hurwitz kill a screech owl with his bare hands? Is this a ridiculous question? Is it wrong to single him out? It is "just" a screech owl. Could he kill one with his bare hands and feel the beauty of the muscles and the potential of the breath? Could he kill five, or ten, the way we used to kill chickens on our farm?

Our system extends this power to people. Through the power of money, one can kill massive numbers of beings one does not even know. What if Charles Hurwitz camped with us for a month, and came to know trees that are different from the trees in Houston? Has he had a chance to experience the redwoods for even a week like the vacationers along the Avenue of the Giants? Of course, they are mostly camped within the sound of the freeway. What about on his own land, where it is silent? Has he had a chance to go there and camp for a while, to feel the beauty of its muscles and the potential of its breath?

Could we, as shareholders of any corporation, be asked to get to know what we own? In war, people do not sleep with the enemy. But is this a war? Every shareholder of MAXXAM *could be given a choice, whether to spend a day killing screech owls with bare hands, or a week camping in the grandeur of the forest. As it is, we can buy today and begin killing tomorrow, without ever laying eyes on what we kill, without penalty for causing extinction.*

And what about the people who do our dirty work? The loggers are asked to kill disrespectfully what they do *know. Word leaks out of Scotia that the loggers are losing morale; that they know what they are doing is wrong; that they know the fast pace of destruction is going to ultimately destroy their own community. The people at the mill report that the machinery is inadequately maintained. They fear that they are going to get "dumped." Charles Hurwitz's community is in Houston, far away.*

So what is the answer? I try to spiral higher. I refuse, for my children's sake, to admit that we are doomed by our own creations. What is wrong? How do we, as a nation, regain control of ourselves? Corporations move among us like bubbles of immunity, able to duck out of the consequences of their actions. How do we regain control so that they do not destroy us?

CHAPTER 9 CIRCLING HEADWATERS

"LEW, can you circle once more? I want to get a better shot of that last clearcut," Doug said.

"You all right, Joan?" Lew asked.

"Yeah."

Doug and I went up with a friend and pilot for Lighthawk, an organization of fliers committed to showing the American public what is happening to its forests. As I approached Headwaters Grove, nauseated, using most of my concentration to avoid throwing up in a tiny four-seater plane, disoriented, so that east no longer felt like east, and wondering why eight-foot-high pampas grass no longer grows just in Argentina but here, where there should be redwoods, I asked myself what I would feel like if I were a female marbled murrelet and it were spring, and it were time to lay my egg.

Before murrelets disappear into the canopy of their nest tree, they often circle several times, perhaps to slow their speed after a rapid flight inland from the ocean. I found myself circling faster in my mind, the season propelling me. What if all I found when I came next spring was this barren ground? How many murrelets have circled, as disoriented as I, and wondered where trees that have been here for thousands of years of murrelet existence have gone?

Why was there pampas grass growing where there should be redwoods? Why did the land hold fans to herself, having grabbed whatever is at hand to cover her nakedness? Where was the forest that from the air once looked soft and continuous? Why did the land look so disorientingly like Southern California, home of my birth, which I fled when they carved it up this same way? Why did the vegetation below me look like chaparral when it was the land of rain forest, our *own* North American rain forest, of which only 10 percent is left? Why? Why, as far as my eyes could see over the ridge, did pink plumes of the pampas wave on a moonscape dotted with token patches of residual stands of trees that MAXXAM is now cutting at breakneck speed?

"Lew, one more time."

Around again. I stared at the horizon. Eyes slid too easily. Why? How? I asked myself. How has it happened that two million acres of ancient redwood forest, all that exists in all the world, is gone except for the holdings in a few parks and these remnants still in private hands? With less than 4 percent of ancient redwood trees left, how can environmentalists be blamed for loss of jobs?

In my mind, I invited America to come see, to rent a plane at the Fortuna airport and witness what lies behind that front face that looks so serene and pastoral and pure. I am amazed that it is not forbidden to fly over this holocaust. Though MAXXAM bars the roads on all sides, keeping the public out of miles and miles of desolation, it is still possible for the public to police with its own eyes from the air. This is still America, I thought, as my eyes slid over the horizon, circling and circling, while below me tiny bulldozers were at work, pulling logs that looked like matchsticks into piles that looked lightly tossed.

Those matchsticks grew for a thousand years, two thousand years. I *know* we all use wood. I *know* I type on paper. I *know* that private property is to be respected. I *know* that people need jobs. But what about the oxygen-producing forest mantle of our earth? When Pacific Lumber Company was a respected, enduring company, logging sustainably, breathing was not incompatible with jobs.

The plane went into another turn and my eyes slid along the horizon, bumped over stumps, dizzied on roads that curled at will, heedless of terrain or erosion. In a cut-and-run game, erosion is not a consideration. It is only a consideration for those who intend to stay. Down the slope of the wing, logging trucks were loaded effortlessly. D-9 Caterpillars looked like harmless toys. I kept having to force myself to recognize the scale and that east was east. I kept having to imagine that this was America and not Argentina, and to realize that the land below me was only a few miles from Fortuna.

I felt like covering myself up with the soft fronds of the pampas grass and going to sleep. What was the use? "Why do we need trees?" I found myself asking practically out loud. It reminded me of the time that someone asked me why I cared so much where I lived.

"You're going to have a dishwasher and a refrigerator wherever you live, so why does it matter so much?"

This same person might ask, "You *have* two redwood parks. Isn't that enough?"

People often do not consider that the parks and the reserves and the wilderness areas around our nation *are not just for people*. They are not just for

Sunday afternoons. They are for Mondays and Tuesdays, when people are at work, and they are for nights, when people are asleep. They are for the osprey, who catches a salmon when no person is looking, and takes it to the nest and feeds the salmon to chicks that no human sees. They are for the screech owl that nests in a cavity where no human eye peers nor ever will. They are for the secrecy of eggs that develop outside of human knowing and hatch with the minutest crack, spilling owl into life. They are for the marbled murrelet female who lands on a branch and tests the moss for depth and finds it deep enough. They are for the privacy of little feet, tamping the moss before she lays. They are for the privacy of the silent, mysterious, ghostly egg, speckled as if behind a veil, or in a fog, that she tucks against the bare skin of her body. They are for the privacy of that very skin, bared through eons to receive an egg against the warmth of the blood beneath the skin.

But private property, what about private property? Hurwitz owns that branch. He can cut it down. He *has* to cut it down. This is America. People need to eat. He has to pay his debt.

I looked below me and thought of my son, David, saying, "Please don't go into Headwaters." Political tensions are high. He is afraid I'll be caught. Punishments for trespassing are dealt out with increasingly unpredictable severity. I looked below me at the moonscape that was once ancient forest, and the misdemeanor crime of trespassing seemed as diminutive as the logging trucks and D-9s.

"It takes two hours to cross that clearcut on foot," Doug said. I have walked with him. He walks as fast as I do.

"How do you do it without getting caught?" I asked, envisioning David coming to see me in jail.

"I'm camo'd out. And a human figure gets lost out there. There are all those stumps to hide behind."

"Of course." I had forgotten how big even the stumps must be. From the air it looked like a desert, but even the stumps must be gigantic. And we went into another turn.

CHAPTER 10 FOG-LARKS

I WENT to see a place in the forest where marbled murrelets nest. One plastic orange ribbon tied to a twig near the ground marks where a fragment of eggshell was found. Another marks the landing place of a second fragment. A third tape marks the place where a chick was found two weeks after a windstorm. The body of the chick was quite decomposed, but his feet were still intact. They were webbed. Beautiful to stand in still dark woods, leaning against the side of one massive trunk of a three-trunked redwood and look up to a wide, crooked branch and know with 90 percent certainty that it was home to a pair of murrelets.

Beautiful to look out into the dark woods and think about the mystery of tiny webbed feet, so far from the ocean. In my mind, I fan them out with my fingers and inspect them. Little sea bird, alone on a branch in a storm, riding out rough seas on a ship of redwood, finally blown overboard so that when the parents returned with a silver fish, the nest was just an empty bed of moss ringed by the droppings of their vanished chick.

"Fog-larks," this is what the old-timers called them. They were plentiful, traveling in the cover of fog, safe from peregrines. A downy fog-lark, flight feathers already growing, but unused. I fan its wing feathers in my mind and let it live just a week longer until it drops and flies.

CHAPTER II GOING TO HEADWATERS

WHEN I first began this book, I spent a considerable amount of time conniving how I could write a book on Headwaters without actually trespassing. Never mind that Doug had escorted the staff of "McNeil-Lehrer Newshour," a reporter from *Time*, and scores of other journalists into Headwaters and its surrounding groves; I didn't want to break the law.

After I flew over Headwaters and got a sense of the lawless way it is being stripped, however, my scruples relaxed. I finally broke down, hired a young woman to stay with my children, assured David that I would be safe, and began loading my backpack.

"What time are we going?" I asked Doug the day before our departure.

"Around nine-thirty or ten. Lucy won't be home till after dinner."

"Nine-thirty at night?"

"Yeah."

"Oh."

The next afternoon, it was raining hard and I assumed we weren't going.

"Well, I guess the trip's off," I said optimistically.

"Why?" Doug asked, with absolutely no clue as to the reason.

"It's raining."

"Oh, that doesn't generally matter," he said in an informative voice.

"Oh."

Somehow, though rain has gotten me wet for forty-eight years, on a trip with Doug things would be different. Doug does not get cold or wet, and for a while I thought I wouldn't either. We stopped and picked up my friend Linda, and she drove us to the gate at the end of Newburg Road in Fortuna. As we approached the end of the road, Doug said, "Cut the lights."

Cut the lights? I was startled into an awareness of what I was doing. I looked out the car window at the cozy little houses around me, peaceful ranches with families in them, and I was aware of how I was stepping outside my normal life. I am accustomed to being the family *inside* the ranch house, not part of a group that is cutting the lights, shouldering packs, and arranging a meeting time. I was aware that many protesters take far greater risks than this, but I am me.

We climbed over the gate that marked PL's boundary and started up the road. It had stopped raining and I relaxed a little. Suddenly Doug slowed and said in a low voice, "Now be careful. Around this next corner there *could* be a security guard."

He hadn't mentioned the possibility of security guards before, and I stiffened again. He gave me instructions on what to do if someone came. He said that we should scatter downhill and then, if we got separated, we would hoot like owls to locate one another. Then Doug went ahead and peered around the corner and announced that no one was there. It was the middle of winter and raining. There was only the ghost of a security guard.

We kept walking and after a while Doug said, "Okay, now we'll be okay."

It was after ten o'clock at night. It was overcast, and a full moon shone somewhere behind the clouds. We hiked up a barren road, and all I can say is that the whole scene was very utilitarian. The moon gave us light but without showing us its beauty and without the mystery of shadows. The road was graveled over, hard, very hard, compacted by the trucks. The vegetation was beat. We came to one small stand of ancient forest, but for the most part it was as though we were hiking through someone's bad haircut. What was left was what MAXXAM didn't need. It was depressing . . . heartbreaking.

We kept hiking steadily, and Doug began to tell me stories. It was like *A Thousand and One Nights*. His stories distracted me as we kept hiking up a seven-mile-grade that was interrupted only by road failures.

A ways up the road, I got a feeling that it was going to rain, and I said that I needed to put on my rain pants and a poncho over my pack. So we both did, and no sooner did we finish than it started pouring again. We walked in the rain, on and on, Doug still telling stories as if it were dry and sunny. Finally we reached the top of the ridge, and Doug asked, "Do you want to go to the near edge of Headwaters or the far edge?"

"What's the difference?" I asked.

"The near edge is about five miles, total, and the far edge is about seven and a half."

"Which is more beautiful?"

"The farther edge."

Doug and I crossed over into the Elk River drainage, and the roads were even worse than on the way up. I was beginning to learn that roads are the Achilles' heel of logging operations. Pacific Lumber / MAXXAM has many miles of dirt roads on their land. Gullies destroy a road that has inadequate waterbars; poorly designed culverts "blow out"; erosion of the roadbed can result in massive slides. As we walked, I felt as if we were viewing textbook examples of hasty and sloppy road construction. It was brought home to me that the Headwaters issue is not just one of trees, but of silt and mud . . . and salmon.

With the vegetation and duff of the forest bulldozed under, the soil no longer serves as an absorbent sponge for winter rains, filtering the water before it emerges downslope from clear springs and seeps during the rainless summer months. Instead the rains pour off the slopes in flash floods carrying huge loads of silt and gravel, filling in the rivers' channels and pools. With no channel to contain it, a river wanders, eroding its own banks and adding to its silt load. Last winter the water of Elk River and Salmon Creek, both of which have their pristine headwaters in the Headwaters Grove, giving it its name, flooded, blocking Highway 101 far downslope. Now the soil of Headwaters lies alongside the freeway in a grave that perhaps few people notice. When I see the lifeless mud, I know, where it originates. I know, too, that it takes hundreds if not thousands of years to create just one inch of topsoil.

Only decades ago, the salmon used to move up Elk River to spawn in such numbers that the beating of their bodies over the rocks of the river channel has been described as sounding like a train. Now the salmon are almost completely gone. Their redds, the gravel areas where the females excavated their nests with their tails, have been buried with thick, pasty silt. To survive, the eggs need to be deposited within open spaces between gravel. There are no spaces now. During winter floods, the young salmon have traditionally laid low in the protection of the deep pools, sheltered from predators within these same precious gaps between the gravel.

As I toiled uphill, I thought about a man named Ralph Kraus who called Taxpayers for Headwaters to find out what he could do about conditions in the Elk River Valley. He has lived there for twenty-eight years and, like most of his neighbors, has always pumped his water directly from the river. He is very proficient at estimating how long after it begins to rain he has to pull his pump from the water so that it doesn't wash away. He told me when I went to interview him that he keeps getting caught off guard because, while the river used to take two days to rise after a heavy rain, it now takes two hours. His community is a dustbowl from the silt flooding over their road. Unaccustomed to viewing PL as a bad neighbor after decades of responsible stewardship of its forests, these neighbors are confused and divided as to what is causing the problem. They have not flown over their watershed, and they no longer feel welcome to trespass beyond the locked gates as they once did when the land was owned by the old PL.

Each time we came to a road that was washing out, "wasting" as it is technically called, Doug commented that we would leave by daylight the next morning so he could document each slide. The fact that he has been photographing landslides and wasting roads for six years became grimly real to me as I tried to scrape thick, clinging mud off my boots. Supported by his slide shows, he's been running back to the media and to EPIC (the Environmental Protection Information Center, in Garberville) with evidence. As I hiked, I became increasingly humbled by my awareness of the people of all ages who have dedicated countless hours, many of them for over a decade, to prevent the devastion through which we hiked. My horror at the destruction was mixed with equal awe for the selfless responsibilty of the people trying to prevent it. The problem is indeed real. The response is appropriate. I felt as I walked along with Doug that I was in the presence of true patriotism, but the loyalty is to more than just people, it is to all species.

At last we came to a long downhill. It stopped raining and suddenly the moon came out effortlessly, as if it could have chosen to all along. I looked across at the skyline of trees in front of the moon and realized by the distinctive shapes of their crowns that I was looking at ancient trees. Unlike second-growth trees, that is, the second generation of new forest that has grown after a primeval forest has been cut, ancient trees seem to follow ancient rules, one branch going one way, the next going another, one short, the next one long, vegetation clotted here, another branch bare.

The redwood is a perfect tree for California, so unlike the beautiful rounded maples of Vermont. It is unpredictable, massive, extreme, shaped into poetry by the winds that cross the Pacific from the Orient. As we walked, the moon played its beams through the canopy, teasing and bobbing and making new compositions the whole way down the hill as if it had suddenly become playful. It was a mix of play and religion, as if the moon owned the trees and the trees owned the moon, whereas before, up the road, there was this great separation of moon and landscape as if the moon no longer knew the land and hung out there, veiled and ashamed of us. It was nice to be able to put the hoods of our raincoats back, and we walked down the hill like a homecoming to a place I'd never been.

CHAPTER 12 HEADWATERS GROVE

WE entered the forest and immediately had to use our flashlights. What struck me was the incredible covering of moss and lichen, the density and variety of the understory. Everything was knit together and held. The ground itself was embraced by moss. We hated to sleep anywhere, because everywhere there was moss, and we didn't want to risk scuffing it. But Doug said there was one place that was his sacred sleeping spot. The moss was intact there too, testament to Doug's respect for this forest.

When Doug asked me to go to Headwaters, I wondered if we would take two tents or one. On the one hand, it seemed an extravagant amount of weight to carry two tents. On the other hand, it seemed a bit awkward to sleep in a small backpacking tent with a young man I barely knew. Doug inadvertently answered my question before we left by asking me if I would carry "the" tent, since he had so much photographic equipment. With what I was already coming to recognize as his usual goal-oriented behavior, the tent was not an issue for Doug, simply shelter. Once we were in Headwaters, however, the whole situation became rather comic. We stood in a dark, dripping, ancient forest at three o'clock in the morning. The rain had stopped at least momentarily; however, our clothes were drenched, and in my opinion we and our packs all needed to be under cover. Doug, who almost never sleeps indoors, said, "I don't think we need to put up the tent."

"You *don't?*" I asked in disbelief. "I *do.*" I hastily put up the tent myself, and we laid our pads and sleeping bags side by side. We took turns using the tent to change into dry clothes, then climbed in. We completely filled the tent, but it began raining hard, requiring that we hastily pull our wet backpacks in with us, like a big turtle drawing in its limbs and zipping itself shut. We lay in the darkness, laughing at the comic moments of the first of what have now been many adventures together, until I finally said, "We better go to sleep if we're going to wake up and see Headwaters."

We woke around eight-thirty in the morning. It was raining pretty hard, and it was so cold that I knew I couldn't afford to get my remaining clothes wet until it was time to go home. I waited in the tent until it stopped raining, then I got up and wandered off by myself with my notebook to write.

I have come to Headwaters with Doug. I sit alone to absorb this place. This is the goal of my pilgrimage, to be here in this ancient grandmother absorbing her wisdom. Each tree is different. The longer one spends in the redwoods, the more one sees this. Now I sit in the burned-out remains of a tree that caught my eye from across the creek. She is like an old crone with her children all standing around her—seven of them, born from her blood and now standing as proof of her power and life.

What awes me about Headwaters is the silence and the richness of the understory. It is an amazing tumbling of carbon over millennia—huckleberries, deer fern, duff, downed branches, rotting stumps, salal, lichen, moss, sorrel, mycorrhizal fungi. As I sit studying the opposite hillside from within my tree, I suddenly ask myself, "How could anyone enter this rich, interconnected grove, where so little has moved or changed for thousands of years, where few people have ever walked, and begin clear-cutting?" Really. This is what men do. They walk into a virgin grove, assess, plan, and set to work. I set to work on my own mental exercise of clear-cutting Headwaters. I decide I would take the smaller trees first and fall them uphill. This would make room for the larger, very beautiful lichen-covered tree to my right. This tree is actually already marked with surveyors' tape and flags. It has already been the focus of some past flurry of planning. I look at a trunk and imagine applying the saw to the bark of the tree. Before I do this, of course, I would have to fire up the saw. This alone is unimaginable. I would have to break the silence. How long has Headwaters' silence been unbroken? The life of the planet?

I look at the moss around me, the lichen and salal. Who would I have to become to ruin this? The word "self" comes into my mind. I would have to detach from "self," from self-consciousness, from conscience. And then I begin to think—this place, above all, could be used for the renewal of "self." There are so many jobs that require loss of self as a prerequisite. A pilgrimage to Headwaters could become synonymous with a search for self. One would have to cross the clearcuts choked with pampas grass before arriving at this grove. Silence here would be a requirement. These groves could be treated like the natural cathedrals they are. Originally the cathederals were built to imitate the feeling of the trees in sacred groves. Headwaters could be our first national pilgrimage site.

CHAPTER 13 WITNESS

WE could not stay in Headwaters Grove as long as we would have liked, because I had promised my children I would be home before they went to sleep. Even though it was afternoon, we were freezing. Doug hadn't brought gloves. I had brought my wool mittens, the kind that fold back so that you can have your fingertips exposed. We kept trading them back and forth as we climbed the road out of Headwaters. We had camped near the headwaters of Salmon Creek. Before the creek reached the runoff from the road, it ran absolutely clear. As it came alongside the road, however, it received its first trickle of silt. We continued to follow Salmon Creek and saw where slide after slide fouled its waters.

As we hiked night fell. We topped the crest of the last ridge, and the lights of Fortuna spread out before us. We sat on the ridge eating, trying to keep ourselves warm with food. Then we walked down the road. My joints still haven't recovered from the impact. Hiking on a logging road is not like hiking on any other surface. I would rather hike with a pack on concrete for seven miles. The road is hard, and it's covered with loose, angular gravel fresh from Yager Creek. One thing that I've noticed again and again is that the rock from these industrial logging operations, which fans out in the watercourses, overwhelming the channels of the creeks and rivers and undercutting the banks of the farmers' fields downstream, is all angular, not rounded like gravel that has been released gradually from the mountainsides by natural forces. It has no wisdom. It is all fresh and naive from the hills. It is washed out en masse and dumped en masse with no history, no heart, and this is what is scooped up and spread on the logging roads. It's the most soulless rock, out of its home. It rolls under one's feet; one fights it to keep from falling. I backpack often, but my joints got so sore that I had to start walking backwards down the hills, slowly inching my way. For the first time since I began backpacking, I became panicked that I might not be able to get myself home. But slowly we made our way down.

In the darkness, we climbed back over the gate where Linda had left us the night before. We had arranged that on return we would call her to get picked up, because we didn't know when we would arrive. We planned to use the phone at Newburg Park, a mile or so down the lonely country road. We hiked past the ranch houses we had seen on the way in, entertaining ourselves with comical explanations for our late-night presence in the peaceful community and our large, poncho-covered backpacks. We came to Newburg Park, where I have come numerous times for David's soccer jamborees, milling with a mass of middle-class mothers and fathers from all over Humboldt County, while our children in various colored shirts had their pictures taken with their teams. The phone did not work. The bathrooms were locked. The playing fields were empty of children. We were simply suspicious people, unable to call, without transportation.

We hid our packs and hiked into Fortuna, to the market where I used to shop when I lived nearby in Ferndale. I walked in, and the checker smiled at me before she had a chance to size up my muddy, sopping-wet clothes. I asked if I could use the phone, and she said yes, the strength of my years of shopping overriding my appearance. After I called, Doug and I went outside to sit on the bench in front of the market. We had eleven cents and a few cashews. We sat sharing the cashews while people pulled up in their cars to make those last-minute purchases before bedtime. After a split-second glance at our muddy boots and wet clothes, they carefully averted their eyes from us altogether, leaving us in the privacy of the disenfranchised.

Finally Linda came. She told us that she had made dinner for us. I was a little concerned because, even though we are real old friends, I had sensed when I first called to ask for Linda's help that her husband, Will, disapproved. Was this simply my own paranoia? They are both public school teachers in a county where tax on "timber dollars" supports the schools. Even among the general population, it is easier to ignore the liquidation of the forests than risk an opinion. When we arrived after the hike, Will was reading a story to their son and didn't even look up as we passed the bedroom. We sat in the living room by the fire, trying to get warm, and I felt what was becoming increasingly familiar dread that my stance on Headwaters might alienate me from many of my dearest friends.

When he finished reading the story, Will came out, and I introduced Doug. From the moment he laid eyes on Doug's face, which, flushed from the hike, appeared as clear and pure as a glass of creek water from the highest headwaters, he was immediately engaged. He wanted to know our whole story, so we sat there eating, surrounded by this warm family, telling what we had seen. As we left, Will said, "I want to thank you for bearing witness to this." I was touched. All of the pressure that I had been feeling was released, and I thought to myself, if I do this right, he won't be the last person who will thank us, and who will particularly thank Doug for six years of bearing witness.

Licorice fern, banana slug

CHAPTER 14 LOSS

PERHAPS I should stop for a moment and tell you a little about Doug's and my origins. Sometimes I look over at this young man, almost half my age, whose path in life has crossed with mine in order to create this book, and I think about what we have in common. Someone said to me once, "Scratch an environmentalist and you'll find loss." While I object to the word "environmentalist" and feel the word "citizen" ought to be adequate, in this dark age when some people do not yet acknowledge that we are all dependent on "the environment," it is still a useful label. I look at Doug and recognize my own childhood loss in his eyes. We both grew up in places that were under heavy pressure from development. Perhaps because we both have a passion to be out-of-doors and because we both love animals, that loss has fueled our life's work, independently, and now together.

Doug grew up in Texas. The other night he came over for dinner and began telling David and me about the pets he had as a child.

"What kind of animals did you have?" David asked.

"Possums, a baby squirrel, a gopher tortoise, lots of snakes: reticulated pythons, Burmese pythons, red-tailed boas, diamondback water snakes, speckled king snakes, gray-banded king snakes, corn snakes. . . . Oh, and I had lots of turtles: red-eared turtles, painted turtles, three-toed box turtles, common snapper. . . ."

I listened enviously, on behalf of David, to the lushness of Doug's carefree childhood.

"Oh, and one time I had a beaver . . . but it was only a guest."

"A guest?" I double-checked, smiling at Doug's choice of words, imagining him giving the beaver a little guest towel.

"Yeah, it gnawed on the woodwork, and my mother told me I wasn't keeping any beavers."

Doug saw the wild country he roamed as a boy disappear beneath pavement, and then he came to Arcata to attend Humboldt State to major in nature photography.

"I was working in a Volvo dealership washing cars a few feet from Highway 101, and I kept seeing all of these gigantic logs going by on trucks," he told me one day shortly after I'd met him.

"I asked someone, 'Where are all these logs coming from?' and no one knew! So I set out to find out, and someone told me that only 3 percent of the redwoods were protected in parks; another person told me about Headwaters, and I went out there to see what was going on. It was so beautiful, I thought, 'If someone could just show people the beauty of this forest and its destruction, they wouldn't keep cutting it down.' So, eventually, I left school, and I went to work photographing and showing my slides."

This is an understatement. As I got to know Doug I was repeatedly surprised by the extent of this young man's influence. When we met he had shown his slides more than 250 times around the country. One night I did some photocopying for him of newspaper and magazine clippings about his work, and I was astonished to find an article about him in *Time*, and photographs in *People*, *Glamour*, and *Sierra*, among countless other magazines and newspapers with trees and clearcuts blanketing their front pages in color.

"I'd only been showing my slides for a few months, traveling all over the country by Greyhound bus, when I got a letter from Pacific Lumber Company's lawyer tellin' me that they were goin' to sue me if I didn't turn over all of my photographs of their land and leave off doing my slide shows. I thought to myself, 'Turn 'em over, my ass.' I was getting ready to do another show that week. No way was I going to turn over my slides."

Doug's mix of formality and Texas colloquialism is part of his charm. One day I said to Doug, "I know this is a stupid question, but have you *seen* old-growth trees cut down?" I thought this question might cause him to throw up his hands and resolve to politely let me off the hook while he found another writer. It was like asking a general in a battle if he had actually seen dead bodies. I expected him to ask, in an exasperated tone, "What do you think this is, an Arbor Day celebration?" Earthfirst! activists were out in the forest hanging a hundred feet up in space from the sides of redwoods with three-hundred-foot trees falling on all sides, and I was asking if he'd ever seen an old-growth redwood be cut down.

"I try to be somewhere else," he said quietly. "I don't like to see it so I try to be somewhere else. But I did see one tree cut down where I think I might have been able to temporarily stop the cutting. We were hiking into All Species Grove, and we heard the saws. I was down a ravine, and they were just on the other side. I got into a spot between two trees where I could watch. Then I got this real sick feeling in my stomach like I was watchin' a woman be raped.

"It didn't sit easy with me at all just watching while this tree was being cut down," Doug went on. "I kept thinking if only I could stall the cutting so the tree might stand for just twenty minutes more, it'd be worth it. It's been standing there for thousands of years, through all those storms . . . but I felt dazed by what I was seeing and I just watched while it fell to the ground.

"Whenever I pass that one tree that used to be standin' there, I have this guilt feeling in me. I'm left kind of speechless, like I don't know what to say to the tree, having sat there and just watched it happen. Every time I pass the stump, I touch it, put my hands on it, and sit down for a while and renew my vows to the forest."

At first I was surprised by Doug's concern for individual trees. For him, each is sacred. At times I have felt a little coarse because I keep being distracted from the forest by the mud, the overwhelming acres of pampas grass, the coffee-colored creeks that no salmon in its right mind would enter. Doug has seen grove after grove that he has known and walked and slept in for days at a time, vanish into rubble, yet every tree that falls is still a singular tragedy.

Doug has unfinished business with the forest. For him there are a few last groves that might still be saved. Finally, enough people are listening that there might be hope. But for me it is all new. I am still reeling with the shock of the destruction, while I am still scrambling to comprehend the crime.

CHAPTER 15 MY OWN BACKGROUND

I GREW up in Southern California, dismayed by bulldozers that destroyed orange groves, forest fires that regularly swept across the front face of the San Gabriels, and smog that could erase the entire mountain range in an afternoon. I was inspired to love nature by a father who sought stability in the wide spaces of the Mojave Desert, trained in the craft of writing by a mother who had a job correcting high school English papers forty hours a week in our living room, and inoculated with an interest in forestry by my older brother, who spent much of his young adulthood hiking the San Gabriels. My sister, who is seven years older than I am, was my source of unconditional love.

When I was a little girl, I waited anxiously between vacations for my brother to come home from college, where he was studying forestry. No sooner did he say hello to the family than he would jump in his Volkswagen and head for the mountains. I did my best to accompany him. As we rode along in the car, he taught me the names of the trees.

"What tree's that?" he would ask, pointing at a Douglas fir.

"*Pseudotsuga menzizii*," I would fire off as quickly as possible.

"What tree's that?" he'd demand, pointing at the closly related big-cone spruce.

"*Pseudotsuga macrocarpa*."

One day he asked if I would like to grow some native California trees. He helped me write to the forest service to ask for seeds. A box full of about twenty envelopes all carefully labeled with common names, scientific names, and directions for planting arrived through the mail. My brother and I built a seed bed, mixed the soil, and planted the seeds. I still remember a grand fir that grew in my mother's yard by the faucet for about twenty years, and two Monterey pines that shot up, crowding out her own plants until she finally had them cut down when they were about thirty feet tall and their trunks each a foot in diameter.

When I was still in college, my sister moved to France, and my brother went to work for Weyerhaeuser Lumber Company, from which he recently retired after twenty-five years as an executive for the Department of Environment.

As my brother became more and more established, my father became the uncontended black sheep of the family. He was obsessed with photography, taking pictures of everything from snails in three-sided, aluminum-foil-lined cardboard boxes, to ant wars, to nocturnal animals that visited our remote camps. One day he came home in a used pink Edsel he had just bought to replace the '53 Lincoln that he seemed to be perpetually backing through the Mojave as he got out of tight spots on narrow, sandy roads.

While I caught the mystery of what I thought my father was seeing when he stared out at the horizon in the Mojave, my brother perhaps saw only a man who was out of work, taking drugs, and neglecting the more practical needs of his family. Thus, a philosophical split began as my brother rose through the ranks of Weyerhaeuser, and I pursued my lifelong passion for writing and drawing inspired by a deep love for nature. As a defense against extinction, I learned to keep the natural world safe on paper. To earn money, I learned to teach children to draw. To this day, I am happiest, pencil in hand, filling in the downy lining of a green-winged teal's nest, tucking beneath that softness a wealth of cream-colored eggs, or helping a child draw a mountain lion, showing him how to shape the head so that it mirrors his own burgeoning sense of power.

When I was nineteen, my father died, and my twenty-four-year-old brother took it upon himself to help shape my future. Thus, at an early age, I lost my spiritual father and gained a pragmatic one with environmental politics considerably different from my own. I eventually became distanced from my brother while I shaped my own character as an artist and writer.

After college, I moved to Mexico, New York City, and finally Vermont. After a childhood in Southern California, I wanted to live in a place that was not enveloped by smog and consumed by developers. I married, raised a stepson, Jesse, and when he was thirteen, gave birth to Suzanne, and three years later, to David. My former husband and I owned and worked a 136-acre farm for almost fifteen years, raising all of our own food plus market-gardening crops such as salad greens, strawberries, and raspberries, and raising sheep and chickens for sale. But I always missed California. I missed wildness—the smell of sage and the sight of the sun setting into the ocean. The stability that was the East's charm, for me became too still. I learned that when one has truly understood California, even landslides and earthquakes are part of one's character.

Now my children and I are back, living near Arcata, home of Humboldt State University, where my brother studied forestry. My brother and I have become closer, but when I found myself getting steadily involved with the writing of this book, I realized that another wedge might be in the making.

During one phone conversation, I made the mistake of beginning to describe my work. After about a half hour of conversation that was going nowhere, my brother, an incurable devil's advocate, had worn me down with rebuttals. I imagined my mother at family gatherings, listening to us disagree, troubled by my pending self-inflicted alienation. Suddenly I said quietly, "I don't think I can write this book after all."

"What?" he asked sharply.

"I don't think I can write this book after all," I repeated.

"Why?" he asked, his tone abruptly softened by my sudden collapse of resolve.

"Because I can't endure the rift it will cause in our family."

My brother paused, then he said, very sincerely, "I would be very disappointed if you *didn't* write this book."

CHAPTER 16 THE THIRD FERN

SOMETIMES I wonder what the world would be like if everyone could draw. Instead of spreading out over the landscape, one would have the option to go inward on a journey. When I want to either buy something or travel, feeling like a new possession or a new place will make just the difference I need in my life, I have found that if I hang on to my seat and begin a drawing, all I really need for fulfillment is a piece of paper and a pencil.

Sometimes when I am in the redwoods, I dream of teaching Charles Hurwitz to draw.

Usually when I first teach someone, they take on too much.

"I think I will draw my favorite cup and saucer. It was my grandmother's. She gave it to me before she died," the person may say. I look at the ornate handle with its extra bends and embellishments and the fluted saucer with its gold rim, and I say, "You might want to choose something simpler."

"Oh, no. I think I can do this."

I concede. I know that life itself will be the teacher, that it will fill the student with awe for the simple arc and ellipse.

Half an hour later, the student says, "I think I will choose something simpler," immersed, not seeing it as defeat, but having discovered all there is to explore in a simple mug.

I would like to take Charles Hurwitz out to Headwaters to draw. People often want to draw a large sweep of forest, but here, too, I suggest that they start with a narrower focus. Just a frond or two of licorice fern growing out of a downed log is enough. There is the arc of the stem to establish, to see its direction through space. It is a matter of vertical and horizontal, and the tracing of a line somewhere in between; 360 degrees, like the hands of a clock. Usually one can narrow it down to range of 45. But where precisely does the center stem cross through the reality of the page?

And then think of the divisions of the fronds. Yesterday I looked at a licorice fern, and I imagined Charles Hurwitz pleasantly lost in the negative space between the positive shapes of the green frond, wanting to do those shapes justice out of a growing appreciation for their beauty. On paper, the space between two fronds can take on the shape of a fern as well. It is here, in this third fern, that the spirit is released. One begins to realize how much has not yet been seen.

CHAPTER 17 HISTORY OF A REDWOOD

I STAND before a single giant, trying to force my mind to comprehend just how long ago I could have come to this very spot and found this very tree, not twenty yards to the left or right, but right here, rooted.

I have been living among the redwoods for almost a decade, yet still I find it hard to comprehend the reality of an individual tree. A giant tree, twelve feet in diameter, may be 1,500 years old. Some may be up to 2,000 years old. I try to imagine the tree as a seedling . . . 1,500 years. . . . It is a journey back in time, an exercise in humility.

I vow to stand rooted until I can feel the magnitude of time embodied in the tree's moss-patinaed trunk that is cracked and ridged from expansion, gnarled, burled, massaged over eons by forces that dwarf my presence. Dwarf me, I think. Awe me. Shrink me to my rightful scale. Teach me my place in time.

Someone told me once that an old-growth redwood towers over a person in the same proportion that a standard-sized door frame looms over a match held upright on the floor.

High above me I cannot even see into the canopy. I study only the tree's ankles, its knees. Who is up there? I pick up a mysterious lichen, red-orange, one I have never seen before, which has fallen from somewhere high above me. I hold a lettuce-like lungwort, otherworldly nitrogen-fixer, soft, mint green on one side, and on the other, that faced the branch, an alien pattern of extra-terrestrial browns. How little I know about my own Earth.

In 700 A.D., within this vast forest that had already stood for at least twenty million years, stretching nearly two million acres along the coast of what was not yet California, a seed sprouted. It grew, one foot tall, two feet tall, beneath a canopy of ancient giants, in a world the white man had never dreamed of.

As this tree grew, the Roman Empire, overripe, fell, leaving its people disorganized, struggling. Meanwhile, the Arab followers of Mohammed invaded northward, conquering with their superior technology. Technology? In 700 A.D.? What technology?

Stirrups. *The Arabs conquered the Europeans due to their use of . . . stirrups. When the mature old-growth trees alive today were seedlings, stirrups were enough to make or break a civilization. I bow my head with the sheer weight of time.*

700 A.D.	Europeans wrote on papyrus.
700s	The concept of the zero was adopted from the Orient by the people of Europe.
771	Charlemagne established a glimmer of European culture reminiscent of Rome.
800	*The tree was one hundred years old.* A hundred years . . . the tree was young. Yet by present-day timber industry standards, a one-hundred-year-old redwood is old. In order to compete with interest rates on the open stock market, timber industrialists often cut on twenty-, forty-, sixty-year rotations. At sixty years of age, the beautiful, red, rot-resistant heartwood for which the redwoods are famous is only minimally developed.
800s	The Vikings developed the art of tacking, giving them the simple ability to sail against the wind so they could make their raids and escape quickly.
900	*The tree was two hundred years old.*

900s	Horses' hooves were first shod with iron shoes.
999	Norwegians blew off course past Greenland and landed in North America (though this discovery was forgotten for five centuries, until the tree was eight hundred years old).
1000	*The tree was three hundred years old.*
1000	Feudalism, with its knights and castles, dominated Europe.
1040	Development of a magnetic needle led to the invention of the compass and eventually to the exploration of the New World.
1096	The Crusades began.
1100	*The tree was four hundred years old.*
1100s	The Chinese experimented with gunpowder, eventually figuring out how to use a small explosion to propel a ball out of a pipe encased in wood.
1200	*The tree was five hundred years old.*
1215	The Magna Carta was signed.
1254	Marco Polo traveled to the empire of Kublai Khan.
1300	*The tree was six hundred years old.*
1324	The first reference to a gun is made in Europe.
1300s	Development of the first mechanical clock, the first primitive eyeglasses, and the first hinged rudder.
1400	*The tree was seven hundred years old.*
1400s	A sailing ship, the caravel, was built with adequate cargo space and strength to cross the Atlantic.
1492	Columbus arrived in America.
1500	*The tree was eight hundred years old and in its prime.*
1521	Cortes conquered the Aztecs in Mexico
1600	*The tree was nine hundred years old.*
1600s	The first European colonists arrived in North America.
1630s	Galileo disturbed the Christian world by declaring that the Earth and the other planets traveled around the sun and not the Pope.
1700	*The tree was a thousand years old.*

1700s The steam engine was developed.
1776 The Declaration of Independence created the United States of America.
1776 The redwoods are seen for the first time by Europeans, the Spanish moving northward to claim California as their own.
1800 *The tree was 1,100 years old.*
1850s Most of the forests of the eastern U. S. were cut. Eastern loggers began moving west.
1860 The internal combustion engine was invented.
1800s Native Americans of the land that was eventually to be called Humboldt County were massacred by whites.
1869 Pacific Lumber Company was founded.
1881 The steam donkey was invented, causing much greater damage to the forests than oxen.
1880s Railroad lines were established in California.
1900 *The tree was 1,200 years old.*
1908 The first Model T was sold.
1914 The first railroad line was built between Humboldt County and San Francisco Bay.
1923 The U. S. government established a policy prohibiting Native Americans from practicing their tribal religions.
1930s The internal combustion engine invaded the redwoods. Caterpillars cut roads to logging sites, gouging the land as it had never been disturbed before, compacting the soil and opening up new sites which had previously been considered unloggable. Chain saws replaced axes and handsaws. Logging trucks transported logs with unprecedented speed.
1960s The ancient redwoods were reduced to a few fragmentary stands.
1985 Charles Hurwitz bought out Pacific Lumber Company.

When will we finally recognize the trees themselves as treasures? When will we look at the map of the United States, see how the ancient forests have been destroyed, and declare as one people that not one more of our heritage trees will be cut? When will we no longer see a giant like this one as board feet, or only as habitat for endangered species, but as endangered itself? When will we see this incredible reach back into time standing right here in our midst and acknowledge, "We have no castles, few cathedrals—we have something better. We still have some of the few ancient forests left on Earth." Not one more ancient tree! Cut. Anywhere in the United States. Not one more ancient snag toppled. Not one more ancient log dragged from its resting place in the forest.

I wait at a stoplight in downtown Eureka, the radio playing, the children talking, while I stare at the butt end of a gigantic log on the logging truck ahead of my car. In the center of the log, deep in the fine grain of annual rings formed 1,500 years ago in the shade of the ancient forest, are the Europeans . . . writing on papyrus. The Vikings . . . tacking, attacking . . . Horses' hooves first shod with iron shoes . . . The Norwegians blown off course past Greenland . . . The compass invented . . . The light turns green and the log moves on toward Scotia.

CHAPTER 18 FUNGI

WHEN MAXXAM's resource manager, Tom Herman, presents a computer-generated video of logging over the next ten years, showing a progression of clearcuts to regrowth to clearcuts to regrowth, the land panting in and out beneath what looks like bouts of mange, he says with genuine glee, "Look! Here it comes again! The orange PL pickup truck! Every time it comes on the screen, something happens!" And, true to his word, the truck zips across the screen and a clearcut appears—tricky, cute, and entertaining. But when someone in the audience asks the serious question, "What about the mycorrhizal fungi?" the question is sidestepped with a joke, as if the concept of fungus were too esoteric to be taken seriously.

Therefore I thought there was something "new" about the concept of mycorrhizal fungi. I thought that word must not be quite "out" from the hallowed halls of science and that biologists should move along in communicating with industry about their important research. I was surprised, however, when I looked at an old botany textbook from 1964 in my personal library and found that it contains the word "mycorrhizae." "Mycorrhizae, literally meaning 'fungus root,' " it says, "are now recognized as organs of fundamental importance in the nutrition of the trees on which they occur."

I picked up a children's book from the same shelf, called *How Plants Work*, a very basic volume that tells only the barest essentials, but there it was again. "These organisms are important in the early stages of a tree's life because they help it to grow. . . . They stimulate root growth and this helps the tree to make more food. . . . Mycorrhizae are important in forestry, especially in poor soils where conifers will grow well with help from these fungi." With the high amounts of rain that we receive in the Pacific Northwest, our soils, particularly those on hillsides, are consistently poor due to leaching. It is only the public's lack of schooling in natural history that allows MAXXAM to get away with minimizing this ancient key to soil health.

But how do the fungi interact with the trees?

If you were to dig up just one of the millions of roots of a single tree in Headwaters and follow it to its furthest tip, you would find that the tip looks slightly swollen and brownish. If you were to take that tip home and look more closely in good light, you would see that it appears furry. Under a microscope, you would see that this "furriness" is not due to hair, but to millions of fungal filaments. In cross-section, you could see that these filaments grow not only outwards into the soil, but also inwards into the host root. Thus the fungus creates a connective layer between tree and soil, extending the roots' reach into the soil and simultaneously increasing the tree's capacity to absorb water and nutrients.

As you might well imagine, this layer is vitally important in the establishment of a seedling. Because a seedling has relatively few, blunt roots, a coating of these filaments vastly decreases the likelihood that the seedling will dry out. In addition, the coating is actually protective, providing a physical barrier that secretes both antibiotics and a gluelike substance that stabilizes the soil around the root and keeps passages open for air and water. In addition, the fungi break down the minerals in the soil, components that can be used by the trees. In a land where summers are rainless and a new seedling may suddenly find itself in parched conditions, its roots are protected by the fungi. Meanwhile, the fungi, which lack chlorophyll and are therefore incapable of photosynthesis, derive sugar from the roots of the seedling. In this harmless, symbiotic exchange, both prosper. It is hard for me, as a parent, not to warm up to these fungi that function so much like foster parents to the seedlings. I am reminded of how, when my children first started school, I hoped that someone would provide the same sort of transitional protection.

When an area is clear-cut and the soil heavily disturbed by bulldozers, however, the host tree species on which the mycorrhizal fungi depend are eliminated. Slash fires, which are a routine practice following clear-cutting, tend to burn hotter and closer to the ground than natural fires, scorching and killing the mycorrhizal spores in the process. Natural fires, in contrast, burn more cooly as they move higher off the ground through the canopy, leaving large logs and whole standing snags behind. Viable seeds and small animals hidden beneath the moisture-retaining logs usually survive natural fires as well.

In an area that is holistically logged through selective thinning, no clear-cutting and no slash burning occur. Most of the animals of the deep forest remain. Redwood stumps and young trees that have been suppressed by lack of light respond to the new openings in the canopy, yet most of the ecosystem remains naturally shaded. When Douglas fir seeds fall from the mature trees and sprout in the newly opened ground, the mycorrhizal fungi are still in place and readily recolonize by spores.

I have come on a walk in the Arcata Community Forest with a man who is a mushroom specialist. This is a place that is walked by hundreds of people every week. Yet walking with this man in a place where I have walked countless times, I see the world turned upside down. Suddenly the important activity of the forest is below my feet. There is a wilderness there, however much its surface may be trodden daily, that is vast and mysterious. This man gently digs into the soil as if he fears he may be prying, and exposes mycorrhizal filaments fanning out in such profusion that it seems as though he has come to the forest ahead of our visit to set up a tricky, exaggerated show of nature's abundance. He tells me a story: "A group of foresters went out in the woods. Some were American and some were Japanese. One Japanese man asked an American, 'Why do you look up when you go into a forest?' The American forester looked at the other American foresters. They were all peering into the treetops or examining the bark. He had to look down to see the Japanese foresters, who were on their hands and knees studying the soil."

He told me one fact that will forever change my concept of the forest. It is simply this: 51 percent of the biomass of an old-growth forest is fungal. More than half of the forest is not made up of branches and needles and the marketable trunks, but of mushrooms and their underground wanderings of filaments. When one walks on land that has been recently clear-cut, the dominant smell is not of sawdust, but of fungi.

CHAPTER 19 FLYING SQUIRRELS

TAKING fungi to heart and cultivating a decentralized vision of them as a great mass of filaments below the ground is a little like the mind shift necessary to perceive that the skin is an organ. We are accustomed to seeing mushrooms and assuming that the filaments that connect them to the earth are their roots. Rather, we must remember that a fungus is actually an underground mass of filaments which periodically sends up aboveground fruiting bodies, the mushrooms, while it also sends out belowground fruiting bodies, the truffles.

Two animals of the forest are particularly dependent on this mysterious mass of filaments that lies beneath the forest floor. Flying squirrels with huge black eyes drift through the darkness of night to the earth, while red-backed

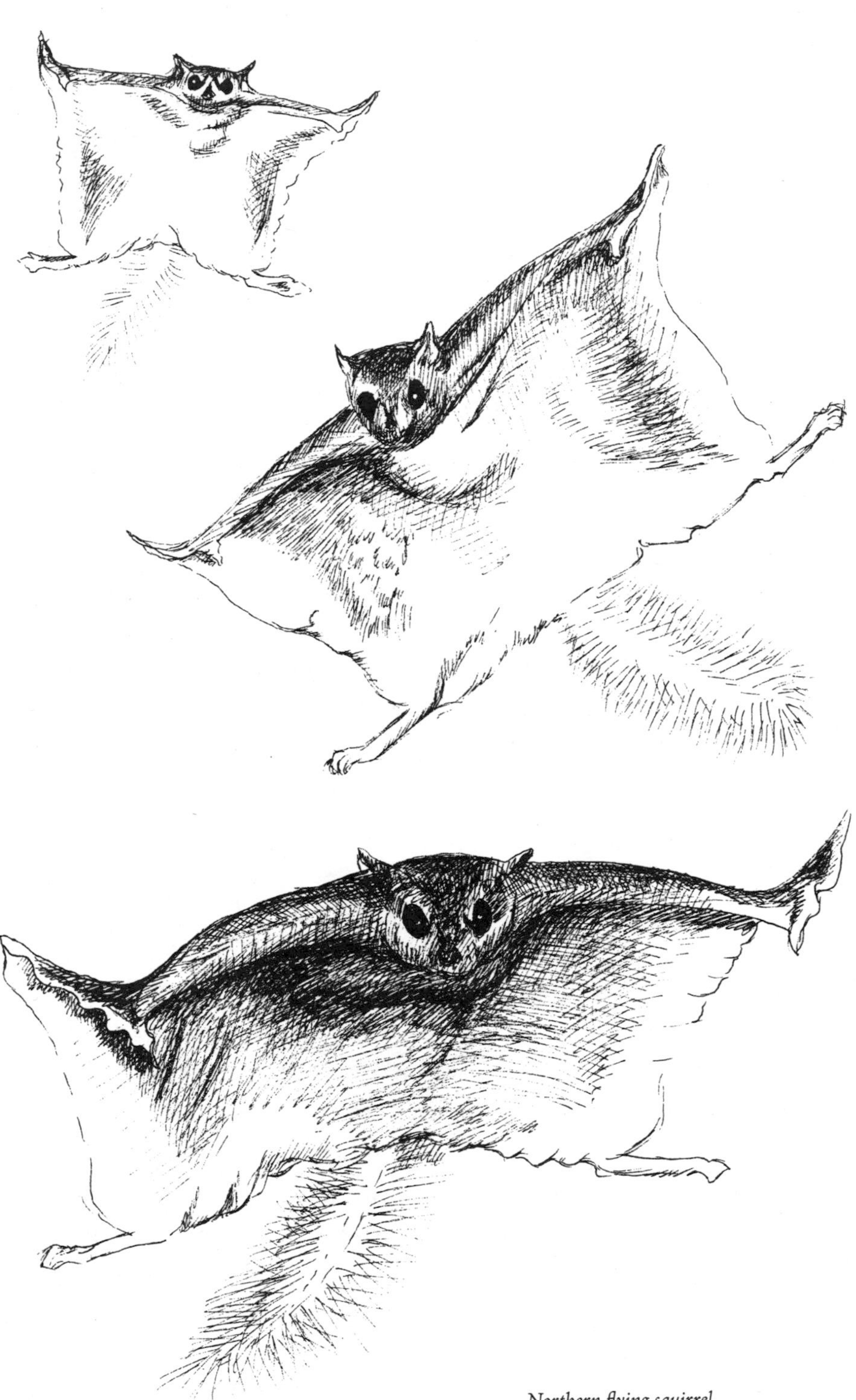

Northern flying squirrel

voles rummage beneath the soil and through decomposing logs, both of them in search of truffles. With such an esoteric mission, they seem the stuff of fairy tales. Yet just as the mycorrhizae are vitally important in the life of the trees, and the trees vitally important in the life of the fungi, the flying squirrels and the red-backed voles are vitally important in the lives of both. The neatness of the triangular relationship that exists between trees, animals, and soil is one of my favorite aspects of redwood/Douglas fir ecology.

I think that flying squirrels are one of the most beautiful species of animals in North America. Unlike most fur-bearing animals, they do not have long guard hairs intermingled with the shorter underfur, so their fur is noticeably glossy. Between each tiny wrist of a flying squirrel's foreleg and the dainty ankle of the back leg stretches a delicate, fur-covered membrane edged with a thin, tan line. This stripe marks the boundary between the rich, reddish fur above and the light underbelly of the squirrel. When the squirrel is sitting at rest, this loose fold of skin with its delicate edging appears like a tailored cape. The squirrel's tail is flat and wide and furred like the body. But most beautiful of all are the flying squirrel's eyes, moist and black and huge in comparison to the tiny head, because unlike the squirrels we normally see, the flying squirrel is nocturnal.

One only has to imagine the forest from the flying squirrel's perspective to understand how clear-cutting and salvage logging are disastrous to the health of our forests. By definition, salvage logging allows the removal of downed logs and dead or dying trees from a forest. Mature trees are more likely to contain insects and are easier for woodpeckers to excavate, providing cavities for numerous animals of the forest, including the flying squirrel. Large fallen trees are vital for protection from predators. Downed logs retain up to 25 percent more moisture than the surrounding forest. Without downed logs, seeds and truffles are apt to dry out, and the entire habitat is a more exposed and inhospitable place for all the small creatures of the forest. When spotted owls, great horned owls, and screech owls are flying on silent wings, and bobcats and foxes are on the prowl looking for food for their young, the underside of a large log near a female flying squirrel's nest tree provides vital safety. From that vantage point, the issue of salvage logging looks vastly different than it does from six feet off the carpeted floor of an office in the state capitol building.

I can imagine nothing more delicious than being a flying squirrel baby. Snug inside its cavity by day, in a nest padded with moss and lichen, the baby flying squirrel sleeps pressed against its tiny brothers and sisters, enfolded by the silky loose skin of its mother. At night its mother steals out the door of the cavity, peers both ways for owls, chooses a landing spot, and silently glides from the nest. Ancient forests and mature forests that have been moderately thinned provide the space flying squirrels need to maneuver. Before the squirrel jumps into midair, she sights her landing place, then leaps, extending her forelegs and using them to steer as she floats through the air. Her flat tail is a stabilizer, tipping from side to side, lifting or lowering to control her course. She goes to the forest floor to feed on truffles.

In the darkness she dashes beneath a fallen log. There she scratches among the needles and leaves, sniffs, catches the scent of mature truffles, and digs into the soil with with her tiny front paws. Her wide black eyes dart all around her as she works, watching for bobcats or foxes that might be alerted to her movement. At last she finds a truffle. It is small, somewhat bigger than a marble. On the outside it is brownish gray, and inside it is filled with dark spore-bearing tissue that provides the squirrel with protein. Over the next few days, the spores of the truffle will pass through the digestive tract of the squirrel and subsequently be redistributed in the forest.

The more the cycles within cycles of ever-increasing intricacy and interconnection that characterize a healthy forest have become understood, the more it has become obvious that a downed log or a snag is not to be valued simply in recoverable board feet. One learns from the forest itself and its creatures that a rotting log's value is beyond the scope of human economy. Rather than even try to calculate, it seems more concientious to simply reach into the rich world it shelters, lift up a handful of soil, and inhale deeply.

CHAPTER 20 OWL CREEK GROVE

ONE morning Doug called up and suggested it was time to visit another grove. I was hesitant. My knees still hadn't recovered from our trip to Headwaters.

"Owl Creek is an easy hike," he informed me. "You can bring Suzanne and David."

"Bring my children?"

"Sure. It's just an easy hike down a ranch road. We'll camp in Boot Jack Prairie and the next day we'll hike over to the other half of the grove."

I wasn't too sure about bringing Suzanne and David, but then another friend, Ken Jorgensen, said that he'd like to come along. He has lived for years in the Cummings Creek watershed adjoining PL land.

"Doug said we could bring children," I ventured.

"Oh, good. I'll bring Gina and Kayla," he said without a second thought.

Suddenly reassured, I asked my children if they wanted to come along. They both said yes. When the night of our departure arrived, Suzanne was getting over a cold, so she stayed home with Ken's wife, Maria.

Maria left us off in the moonlight up on a ridge above Boot Jack Prairie. Boot Jack Prairie is one of the last upland prairies in the world totally surrounded by ancient redwoods. It has a luxurious feel, sweeping across the south face of a mountain that is buffered from any road noise by a series of ridges rising out of the valley of the Van Duzen River. The prairie covers an old slide. One imagines that it might originally have been vegetated only by native grasses, but now European grasses, spread by the cattle that periodically graze there, have become intermixed. I know Doug cannot look at the prairie without watching for the bobcat he saw dash across it one day, or the pair of bears that he saw ambling lazily into the sunset one evening. I will always remember my first dawn there, listening to the varied thrushes claim the forest below as their own.

After a night on the prairie, we entered the thrushes' forest, and the atmosphere around us changed instantly. We made our way down a small creek bed, through mixed-age trees and again, as in Headwaters, the understory impressed me as much as the trees themselves. Any ancient forest is a whole ecosystem, not merely trees. This is why the bumper sticker so common in this county, that reads, "Trees, America's Renewable Resource," is so misleading. We must ask ourselves not if trees are renewable, but if our *forests* are renewable when the trees are clear-cut on fifty- to sixty-year rotations, which is common practice in liquidation logging. Most experts think not. Without preservation of the diversity that characterizes an old-growth forest, the fabric of the whole ecosystem begins to unravel as has occurred in Europe, where replanted forests are not surviving past the second or third generation. At one point we came to a strange stump which must have been sawed at least a hundred years ago. We pondered by whom, as we poked around in the deep duff covering its top. The stump proved to be too decomposed to show rings. Instead we found a California slender salamander, only a few inches long, curled up like a miniature rattler.

Owl Creek Grove has been severely diminished in size over the past few years. It is one of the highest-elevation groves, and biologists feel that if there is to be replanting of high-elevation redwood forests, seedlings from this grove or the few groves like it will have to be used. Lowland redwood seedlings would not be adapted to the uniquely harsh and dry conditions.

We punched out of the forest onto a landing where logs were piled during the last logging "show." The silt in the puddles hung suspended.

"A truck passed by here not long ago," Doug said. I stiffened, checking David's whereabouts.

"If you hear a truck . . ." Doug began.

"Oh no," I said. "Here we go again." I couldn't imagine successfully scurrying all of the children off the road like a mother hen.

"No, really, listen up," Doug continued. "If anyone comes, go *downhill*, not up. You can't run fast enough uphill." The children drew closer, looking left and right.

"If we get separated, hoot like owls." At this the children got lost in the fun of practicing owl hoots.

"If you get caught, don't mention my name. They've been waiting for me for years," Doug continued, eliciting a whole new stream of jokes.

But privately, I imagined Doug's reality. He is more like an owl than a young man. Like the owls, he lives as a fugitive when he is on Pacific Lumber land. He has few rights. He has chosen this situation; the owls didn't. Yet in a way, like me, he didn't choose it. It chose him. The forest has chosen all of us who are involved in its protection. It has tapped us on the shoulder, tapped our souls and engaged us. When we respond, we jeopardize ourselves, throw in our lot with the owls. A friend of mine, Pat Harris, said to me when I expressed my fear of speaking up on behalf of the forest. "I tell my children, don't ever *not* do the right thing because of fear."

As another friend's daughter had printed on her high school graduation announcement, "Life is not the absence of fear, but the affirmation of life in spite of fear." Perhaps fear is too strong a word to explain why people are reluctant to get involved with controversial environmental issues like protecting the redwoods. They simply don't want to be viewed as radical. I want nothing more than to dissolve the polarity that plagues this county and this country, to bring us all back to center—the owls and the pussycats, the loggers and the environmentalists, the business community, everyone—to put us all in the same life raft, which is our Earth. We are sailing out to sea in our own universe, whether we acknowledge this or not.

We walked a long way on the logging road and then left it and bushwhacked across recent clearcuts, finally hooking up with another logging road. We arrived at what was once another part of Owl Creek Grove, now cut down. Feeling dismal, we dropped our packs and sat surveying the destruction. Stumps towered over our heads on humps of soil eight feet in the air where the soil had been bulldozed away to make fall beds. The forest floor was so utterly stripped and rearranged, pushed this way and that, that it was hard to imagine that it had so recently been clothed in the rich understory of the undisturbed forest through which we had just walked.

As I looked at the stumps, I had an irrational, shameful feeling: in the way that one can look at a child victim of incest and associate the victim with the crime, the stumps and the soil seemed "hard," worldly-wise, as if they had somehow brought this tragedy upon themselves. It was a cheap way for me to disassociate from what was around me. Like a mother drawing her own precious child closer to her in the midst of children who have been neglected and abused, I had trouble seeing the stumps for what they were. Though the trees themselves are gone, the stumps are no less innocent, no less alive. They never asked to have men arrive and eye them up and down and return with equipment to destroy them. They simply stood and then the trees above them fell.

Doug pointed across a small tributary of Owl Creek and said, "That's where the tree-sitters were."

"*Right* there, across the creek?" I asked. He referred to Tim Ream, the fellow I heard speak at the rally in Arcata, and the other Earth First!ers who had protested this logging by bodily occupying the grove.

"Yeah, just on the other side."

I felt called to go see the stumps and set off with David to cross the creek and ascend the slope. The creek itself was covered over with redwood tops. We could hear the water, but the creek was no longer visible. Logs, branches, and twigs were strewn every which way, foliage still fresh. We crossed carefully on the mat of logs, never once slipping or falling through because the layer of debris was so thick. The trees right along the creek had all been cut down. There was no riparian corridor left, that is, the vegetation that lines a watercourse, holding the soil and providing shade to maintain cool water temperatures for wildlife such as salmon. It was sobering to traverse destruction left by someone approximately my own age with David, who will inherit the mess. He was traversing the greed and neglect of the previous generation, as he'll have to do one way or another for the rest of his life.

We climbed the slope and sat down and I stared at a nearby stump. I said to David, "Perhaps that is the tree in which Tim was sitting, or perhaps it's the tree he watched being cut down."

I picked up a small green spray of redwood and stared at it. The ground was littered with small branches like it. There are two types of needles on an old-growth redwood. On the lower branches, the needles are flat and arranged fairly conventionally on the stem. But the upper canopy of a tall redwood lives in an entirely different climate. It is exposed to wind and sun that do not penetrate to the understory. The needles in this dessicating environment are tougher, smaller, more closely grouped, and they do not resemble the redwood twigs one normally sees on the ground.

These small branches were everywhere, bearing witness to the recency of the cut and the height of the trees that were taken. The branches could have been the hopes of the tree-sitters: tough, fresh, shattered. The police surrounded the trees, and the Cat drivers began bumping the trees with their blades. The tree-sitters finally climbed down and were arrested.

We walked back to the rest of our group and Ken was telling a story about visiting a friend's house, a logger, whose son proudly brought out pictures of himself with the giant logs of old-growth trees he had cut down.

"You know, Joan, it's a status thing to get to cut these trees. For a young guy to get to work logging old-growth, he probably has to know someone. He was showing me pictures he intended to show his grandchildren one day, to show that he was logging when there were still real trees. He was carrying on a family tradition. He was saying, 'Look at them big pumpkins.'"

"Big pumpkins?" I asked.

"Yeah, they call them big pumpkins."

I thought to myself, "This logger's pride in an extremely difficult and dangerous and worthy profession is endangered by the attitude of the company. So much in life is a matter of balance. Without water we die of thirst; with too much we get flooded out. If the forest were managed properly, this man's grandchildren might have selectively cut 'big pumpkins' and no one would have minded."

Gray fox

CHAPTER 21 CHARTRES

I REMEMBER, as a young woman in my twenties, driving across an open plain in France and seeing Chartres Cathedral appear on the horizon, gradually growing in size as my little Citroën Deux Chevaux approached. Walking into Chartres was an unforgettable experience. Power beamed down to me from its windows. It did not care what religion I was. It freely insisted on delivering itself straight to my heart. As a component of that power, there was an element of time—I was receiving something proven. How many souls received instant reassurance from that colored light and from the intricacy of the patterns above, that beamed down all the care that went into their slow creation? If that intricacy bespoke only the glory of humanity, not God, it would still create believers. The colors of the glass etched brightness into my soul that remains as vivid almost thirty years later as it was that day.

I think of the modern excuse for a cathedral, the church with token panes of colored glass that "symbolize" a stained glass window, yet were really cut on a page of calculations by the church finance committee. They have the same effect on the soul as walking through a tree farm. One opens up, anticipating renewal, and feels so little. Those elements of time and care and detail are absent. There are no surprises to sustain the soul.

Walking into Headwaters is like walking into Chartres. The forest is in charge of one's state of mind. The results of thousands of years of evolution envelop one's soul. The ability of the Earth to feed and clothe itself is glorified in greens and browns and reds—mosses, lichens, fungi, sing out the praises of decomposition while the varied thrush, the winter wren, the trillium, the yellow redwood violet, bring one back to the preciousness of the moment.

What if Chartres had been bombed? Destroyed. I could never know completely what I had missed. Pictures, descriptions, could not place me under the beams of that light penetrating through colored panes from forty feet above. It is the same with clearcuts and ancient trees. There is so little relationship between the two that it is hard to make a connection. To stand in the wreckage of Chartres and try to tell someone, "Oh, but these little melted bits of glass, when they were high above me in geometric patterns that spoke of the highest order of the human soul . . . they made indelible imprints on my own soul, they forever elevated my concept of man's potential and my own. . . . " And the person would nod and sympathize, but they would have missed Chartres.

CHAPTER 22 SALMON

WHILE we were hiking through the clearcuts that were once a part of Owl Creek Grove, Doug grew quieter and quieter. He pulled ahead and lagged behind and finally went off altogether, winding down the switchbacks of the logging road like a lost boy. I could see him setting up to photograph the clearcuts, so we left a note and started back. When he caught up with us, he and I stood alone at an overlook. I asked, "Are you sad?"

"Sad?" he asked in a flat voice. A darkness had come over his face, and a heaviness had entered his body that left him silent. He is so accustomed to loss that he does not acknowledge it anymore, it simply enters uninvited. "I guess so . . ." he said.

"I am just trying to understand your feelings. Imagine if our roles were reversed," I said. "Imagine if I had been hiking here for six years and you were just coming here for the first time, encountering this land after it had been clear-cut." He grew reflective and looked off across the ragged hills, mentally erasing groves one by one from his memory as they have been erased from the landscape, severing his connection with his own history.

"Yeah, I guess it would be real different."

"Do you feel hopeless?"

"Hopeless? No. I just want to save what's left."

A shadow, like the shadows of the clouds that kept moving across the barren landscape, had moved across his face and fixed there. Like so many people who have fought this fight at their own expense against well-paid lawyers and modestly paid loggers for ten years, five years, even one year, Doug is tired. He sees that there is so little old-growth left that there's no question that it must simply be preserved.

Yet he is emphatic that what is left must be seen not simply as stands of trees, but in terms of the integrity of the watersheds in which those trees exist. Though they will nourish our souls, we must save more than isolated cathedrals. The six ancient groves left in Headwaters are part of total ecosystem that is still functional. The wild strains of salmon that call Elk River, Salmon Creek, Lawrence Creek, and Yager Creek home are still out there with their insatiable drive to return to their natal creeks. Their capacity to repopulate is as strong as ever. But as abundant and vigorous as the salmon populations have been in the past, they now hang in time as little more than threads. With each season of inhospitable conditions, those threads, those frayed remains of a great rope of life, are being snipped and let go, back into eternity.

Just a hundred years ago, it would have been inconceivable that our culture might ponder whether or not to preserve its salmon. It would be like debating whether we should preserve the air we breathe. Not only were salmon an immensely abundant food source for humans, they played a vital role in the life of the old-growth forest. The muscular tide of salmon that appeared each fall from the vast reaches of the ocean, drawn upstream as if by reverse osmosis, delivered vital nutrients back to the land.

Listen to this. Pause and imagine. Unlike western salmon, the salmon populations on the East Coast do not die when they return to their home streams to spawn. In the West, where soils are thin and easily depleted, their precious nutrients flowing downstream with each storm, the salmon die far inland, in a mysterious gestalt where the land over eons has called back its nutrients and been annually replenished. How does this come about? How, in a nutrient-fragile ecosystem, does it just happen that the salmon are genetically programmed to die upstream after spawning, whereas in an ecosystem with more stable slopes and soils, this does not occur? Ancient systems don't require analysis. They simply work.

Over wordless eons the Earth has taken care of itself, evolving a multifaceted system more intricate than the windows of Chartres—streambanks reeking for miles with the rotting corpses of fish that have spawned and died; an eagle, clasping a dead salmon and carrying it to chicks high above in the gigantic nest, which one day will collapse the tree with its sheer weight, delivering a rich dose of fertilizer for the forest floor; a bear, devouring its fill and heading to its den, pausing to defecate in the woods, dropping the ocean's nutrients far upslope to filter through huckleberry and salal roots to the shallow roots of ancient giants; a kingfisher with an immature salmon in its beak, entering the tunnel of its deep burrow, shuffling with the fish through the darkness to drop it before by its uncounted young, who defecate, scratch down soil from the burrow walls, and eventually grow up, leaving behind a capsule of nutrients to be mined by roots from above. Long before man ever set foot on this continent no more beautiful nor well endowed than any other, this system was fine-tuned. Yet with the fickleness of shoppers debating over a shirt at the mall, we decide whether to let it go.

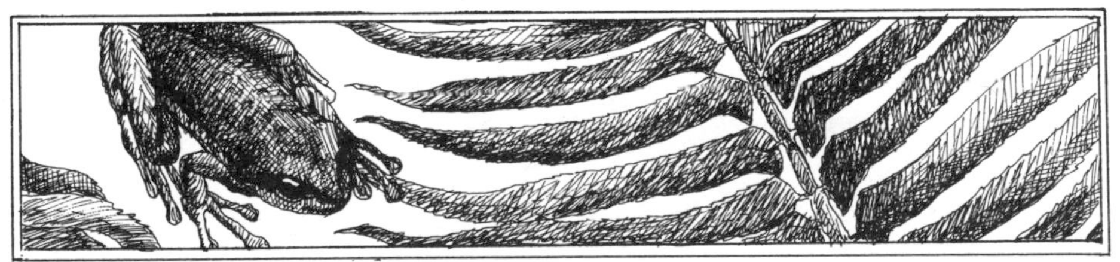

Pacific tree frog, sword fern

CHAPTER 23 TARGETS

LET me pause for a moment and recount some of the events surrounding the illegal logging of the ancient redwood/Douglas fir grove of Owl Creek, second largest grove in the Headwaters complex, and the court cases that resulted. The recent history of Owl Creek Grove is a good example of the forces at play, with only subtle variations, since MAXXAM took over Pacific Lumber in 1985.

Newcomers on the scene of forest politics often feel the way I did once when I attended a "steering committee" meeting of my children's elementary school. I left in despair after half an hour of what sounded like "The NYB depends on the HRF and the FSH can only be decided on the basis of the LQX." To save you from leafing around, as if you were trying to keep track of the characters in a rather sterile Russian novel, let me explain that a THP is a timber harvest plan; CDF is the California Department of Forestry, which is controlled by the California Board of Forestry, which is composed of twelve years' worth of Republican governors' appointees, the majority of whom are sympathetic to the timber industry; and PL is Pacific Lumber, a name which, in this part of the world, is used interchangeably with the name MAXXAM and generally considered to be synonymous, however cruel this may be in light of history. DFG is the California Department of Fish and Game, the state agency charged with looking out for the well-being of California's wild animals, specifically through the enforcement of the state Endangered Species Act. The U. S. Fish and Wildlife Service has a similar function, for animals listed as threatened or endangered nationwide. EPIC is the Environmental Protection Information Center, a citizen organization based sixty miles south of where I live, in the little town of Garberville, California. For twenty years the members of EPIC have used litigation to diligently protect fragments of the last ancient forests that remain on our continent.

On April 11, 1990, Pacific Lumber filed a Timber Harvest Plan to log 237 acres at the heart of Owl Creek Grove. Filing a THP is standard procedure prior to any logging operation. In the old days, THPs used to be only a couple of pages long, and they were virtually assured of automatic approval. In essence they simply said, "Here's what we're going to cut next," to which CDF would respond, in essence, "Why not?"

At the time of filing of the Owl Creek THP, however, the marbled murrelet was just being considered as a "candidate" for protection as a rare, threatened, or endangered species on both a state and a federal level, and as a result, CDF denied Pacific Lumber's THP because PL had not provided adequate information on the marbled murrelet's status in the grove. PL, unaccustomed to being asked for details about the land it intended to log, simply refused to conduct any surveys.

On March 12, 1992, the marbled murrelet was officially listed as a California Endangered Species.

On March 13, the California Board of Forestry overruled CDF's denial of the Owl Creek THP and approved the plan due in large part to Pacific Lumber Company's lobbying of the Board.

At this point, EPIC and the Sierra Club sued the Board of Forestry (*EPIC and Sierra Club* v. *Board of Forestry*), alleging that the THP would violate the California Endangered Species Act. They succeeded in convincing a Humboldt County judge to require the Board of Forestry to reconsider its approval of the plan. On March 16, the Board again approved the plan, but this time on the condition that PL perform the requested marbled murrelet population surveys and "share" the results with the California Department of Fish and Game.

By June, sticking to the "letter" of the Board of Forestry's requirements, PL completed a hasty marbled murrelet survey. Company executives put the results in an envelope and mailed it. Before the letter actually reached its destination, however, Pacific Lumber commenced a surprise, all-out massacre of the grove. Unlike the usual procedure for logging virgin forest, Pacific Lumber scattered numerous timber-fallers throughout the THP area with instructions to drop as many trees as possible. This massacre not only occurred on Sunday, a customary "day of rest" for PL, but on Father's Day.

After three days of illegal logging, which netted the company $1 million in ancient giants, PL was finally stopped by agency intervention. Heated discussions continued all summer and were finally suspended for the Thanksgiving holiday, just after the marbled murrelet had been officially listed as "threatened" on the federal level. On that Wednesday, just before the employees of the U. S. Fish and Wildlife Service and the California Department of Fish and Game went home to relax with their families, their agencies reemphasized that Pacific Lumber *was not to construe that during the discussions they had in any way been granted permission to harvest.*

Pacific Lumber chose not to heed the agencies' warning, however, and over the holiday, in spite of the Endangered Species Act's clear prohibition against injuring or killing listed species, they sent their loggers out to begin a second assault on Owl Creek Grove.

Chuck Powell, the man who took me with his son to see torrent salamanders, spent a few hours the day after Thanksgiving showing a visiting photographer the devastation left by the Father's Day cutting of Owl Creek Grove. They drove out the next ridge to the south of Owl Creek in Chuck's pickup and stopped, expecting to hear silence or perhaps the distant sound of Yager Creek from the valley below. Chuck couldn't believe his ears. He heard the sound of chain saws coming from the direction of the grove. PL was at it again! In just a half hour's time, he heard *eight* old-growth trees fall. He also noticed that, strangely, he did not hear the normal concurrent sound of bulldozers, and suspected numbly that the ancient trees were simply being dropped as rapidly as possible to fragment the grove so that murrelet occupation would be less likely in the future. Chuck and the photographer raced back to telephones to alert government officials and the public that a second massacre was under way.

EPIC's lawyer, Mark Harris, leapt into action, phoning the U. S. Department of Fish and Wildlife, CDF, DFG, and finally even the FBI to inform them of the illegal logging. To give you an idea of the government "apathy" that conveniently gives the timber industry a loose rein, Mark was greeted with a variety of responses, from "That's impossible. It simply isn't happening" to "Is this *really* such a big deal, Mark? It's only some trees, and what about jobs?" Even local judges would not take the suit against PL, so it was moved out of the area to the state Court of Appeals in Sacramento. EPIC finally obtained an emergency stay, and the logging was stopped after days of illegal cutting. Though EPIC sought to prevent PL from actually removing any downed trees, stating that such action would result in even more disturbance to the grove, Pacific Lumber claimed that the trees would rot. Though it takes centuries for a redwood to decay, Pacific Lumber was allowed to drag another $1 million worth of lumber out of the forest. This is but one of countless instances of EPIC and other environmental organizations functioning, at private expense, in place of our government agencies, whose efforts have been sabotaged due to corporate influence.

EPIC at that point decided to file a lawsuit. The suit bore the no-nonsense name of *Marbled Murrelet and EPIC* v. *Pacific Lumber*. Just the name touches me. I think of the marbled murrelets that I have seen bobbing on the ocean, or those that I have watched speeding inland through the fog to a nest tree. I laugh inside to think of a marbled murrelet bringing a lawsuit. And I grow silent with joy because it won.

Mark Harris, at the time, was a fledgling environmental lawyer who launched himself into the case with the resolve of a marbled murrelet chick jumping from its home redwood into thin air. For political reasons, state and federal wildlife officials continued to refrain from taking legal action. Only EPIC, operating on a shoestring budget, stood up for the law of the land. The fact is, the Endangered Species Act clearly states that it is illegal to "take" an endangered species. EPIC contended that eliminating crucial breeding habitat, namely cutting down the very trees to which the birds return each spring to nest, constitutes a "take." In hindsight, now that EPIC has been victorious, this sounds ridiculously obvious, but at the time, prosecuting the case was beyond an act of faith. It simply had to be done.

Mark, a thirty-three-year-old surfer/attorney just three years into law practice, found himself pitted against one of California's oldest and most prestigious law firms, Pillsbury, Madison & Sutro, backed up by Ukiah's Rawles, Hinkle, Carter, Behnke, & Oglesby, and led by Pillsbury partner Alston Kemp, Jr. Mark found himself sleeping nights on the floor of his office, sustained by EPIC staff who brought him food, did his laundry, and helped him process the massive amounts of paperwork being dumped on him by the opposition.

Meanwhile, in Colorado, attorney Macon Cowles, reading about Charles Hurwitz in his morning newspaper, turned to his wife, Regina, and said, "I'm going to sue this son of a bitch."

Cowles had just won a $24 million verdict on an environmental case in Colorado for which he had mortgaged his own house in order to prosecute. He was yet to learn that his role as one of the lead counsel on the *Exxon Valdez* case would help win a $1 billion settlement to clean up environmental damage in Prince William Sound, Alaska. Cowles got in touch with Mark and said that he wanted to see Owl Creek Grove before he agreed to sign on.

So Cowles came west to meet this young man and fly over Owl Creek Grove.

From the air, the grove was an isolated fragment of luxurious forest surrounded by barren, clear-cut slopes and occassional areas of sparse second-growth trees. Cowles was instantly convinced. Before the case was over, he would spend more than $200,000 of his own money in this new gamble on the strength of our legal system. Visiting Philadelphia federal judge Louis C. Bechtle was not to let Cowles, Harris, the American people, and the marbled murrelet down. In fact, Cowles obtained a stuffed murrelet from DFG, which watched the proceedings in the San Francisco courtroom each day through its beady glass eyes and, conversely, was in plain view of the judge as the plaintiff in the case.

The Endangered Species Act is a line drawn in the sand. It is a statement of self-limitation. It is a tool, like a smoke alarm, set to warn us to ask, "Is there enough of an ecosystem left for *us?*" It is not something to dismantle, just as we do not dismantle our own smoke alarms, but something to heed.

As out-of-town attorneys and their assistants met fearful survey biologists in out-of-the way cafes and hammered PL for its records, they began to get a sense of the repression and distortion that have become the norm in Humboldt County. An unflattering picture of PL began to emerge. They discovered that Pacific Lumber had concealed *seventy detections of murrelet presence* in Owl Creek. Judge Bechtle labeled PL's $2-million murrelet and spotted owl surveys as "highly suspect." Despite another $1 million that PL spent on legal defense, Bechtle commented that "EPIC has served the public interest by assisting the interpretation and implementation of the Endangered Species Act."

Judge Bechtle wrote in his final decision that "EPIC has proven, by a preponderance of the evidence, that marbled murrelets are nesting in THP-237. . . . Moreover, Pacific Lumber's . . . marbled murrelet surveys were either designed to fail to detect murrelets, or they were administered with indifference." Judge Bechtle concluded that "a permanent injunction prohibiting Pacific Lumber's implementation of THP-237 is warranted."

The Owl Creek decision made history because it established the precedent that killing the *habitat* of an endangered species constitutes a "take" of the species itself and is therefore illegal. It was also the first time that the Endangered Species Act had been enforced *on private land*.

Pacific Lumber appealed this decision and the case went to the United States Supreme Court, where it was upheld in the spring of 1997.

Lest you have any doubts about PL's feelings about the little glass-eyed plaintiff in this case, in his decision Judge Bechtle observed, "At the end of the 1992 survey season [September] Pacific Lumber's Resource Manager, Thomas Herman, hosted a party at his home for Pacific Lumber's forestry staff, which included the company's marbled murrelet surveyors. . . . At the party, there was a target of a marbled murrelet on a dartboard, at which the attendees were throwing darts."

CHAPTER 24 THE MAHAN PLAQUE

I HAVE been making regular trips to Humboldt Redwoods State Park and Redwood National Park to do my writing. I have a need to let the redwoods speak for themselves, to do their own work. I am like an ant, scurrying around, attempting to give the big picture, and they look down from their own canopies and see miles and millennia. I cannot always go to Headwaters. The locked gates, the long expanses of devastation, and the close cooperation between the Humboldt County Sheriff's Department and PL security discourage the fainthearted.

One evening I stopped in the drizzling rain at Founders' Grove in Humboldt Redwoods. I wanted to visit the Mahan Plaque. I had been to it once before, but long ago. Now that I am becoming embroiled in this fight, I wanted to go back and read it again.

As soon as I stepped from the car, the sound of the freeway nearby said, "This is not Headwaters." Yet, as soon as I crossed the small parking lot, the trees embraced me. In spite of the fact that Founders' Grove is a principal destination for visitors to the redwoods, the trees instantly absorb pettiness. And, best of all, because of the rain, I was alone.

Founders' Tree. The path leads directly to this giant. The sign reads, FOUNDERS' TREE, HEIGHT 346 FEET, DIAMETER 12.7 FEET, CIRCUMFERENCE 40.0 FEET. It was the last statistic that stopped me that evening: HEIGHT TO THE LOWEST LIMB 190.4 FEET. I do not readily think in numbers, so I convert figures like this to skyscrapers. A nineteen-story building could fit under the first branch! I looked up. Nineteen stories. . . . After that, the tree ascends into an immeasurably private world. I wandered back through the grove, eventually encountering small signs that lead me to the Mahan Plaque. It is a small boulder with a bronze plaque affixed to its front.

I have come to visit the Mahan plaque, dedicated to Laura Perrot Mahan and James P. Mahan. I want to brush up against their dignity and their success. It is easy to be dignified when you're dead and no one is afraid of you. It is easy, also, when there are lots of redwoods.

The plaque has greened with a patina of age. I was all alone in the grove, except the Mahans. Mrs. Mahan, in her elegant and tidy dress with its high, lace-edged collar, was positioned behind her husband on the plaque, but her name came first.

Laura Perrott Mahan 1867–1937
James P. Mahan 1867–1937
Pioneers in the Save-the-Redwoods League

They were born in the same year and died in the same year. I was touched when I saw this. The story I have heard is that on November 19, 1924, the Mahans discovered logging beginning in what is now Founders' Grove. Laura lay down in the area where the loggers were working and wouldn't move, while her husband went to Eureka to secure the paperwork to save the trees. This has become one of the primary functions of Earth First!, to hold off actual logging while an organization such as EPIC obtains a temporary restraining order. Next to the plaque, there is the last huge stump, now covered with moss and lichen, where the logging was halted by Laura's brave action. I do a little math, and I realize that the first blockader for redwoods was not "young and irresponsible," as proponents of the timber industry like to characterize their opposition, but a fifty-seven-year-old housewife.

Thousands of tourists visit Founders' Grove every year. If the police or a deputized logger had dragged Mrs. Mahan away from the site and arrested her for trespassing and put her in jail, this grove obviously would not be here.

In the late afternoon twilight, the redwoods dripping all around me, I squat down with my back against the plaque and say thanks. As I pause, staring out into the shadowy trees, I feel that Laura Mahan has laid her hand softly on my back so I can just feel it and be reassured.

On my way back, I took a detour to visit the Dyerville Giant. As the reclined form of the tree became visible through the forest, I thought about the fact that salvage logging has been going on all winter in the rain-soaked groves of Headwaters. Under the state "salvage exemption," PL can move at will through the old-growth forest, without permits, extracting logs like the Dyerville Giant, dragging their nutrients away on skid trails.

I lay my hand on the fibrous flank of the Dyerville Giant. Despite the traffic noise, it is bliss to be alone. Even though the Dyerville Giant lies on its side, it is no less a tree than when it fell. It still emanates stateliness. On its shattered trunk a lone honey mushroom is growing. Then I find a whole cluster of honey mushrooms, pale yellow, darker orange in the middle, glossy in the rain. I feel like the Dyerville Giant has already given birth.

I escape from the rain into a goose pen tree. I guess this tree would be "salvaged" in the Headwaters groves as well. As it is, it is sheltering me. This tree might well live another five hundred years though its core was burned out probably hundreds of years ago by fire. Settlers called these trees goose pens because they used them as pens for their animals. This "pen" might have housed a single rabbit. I can barely squeeze my hips into the cavity and sit down. The trunk curls up and around me so that I am in a throne. The mass of the tree absorbs the sound of the highway behind me, though the engine noise echoes back from the wide trunks of the trees in front of me.

The sorrel that carpets the ground bobs a leaf every now and then when raindrops hit. My soul fights the sound of the trucks. Between trucks this place is perfection, but I count—twenty seconds between trucks, thirty-two, fifteen, fourteen. . . . The longer intervals are almost worse. The soul recovers and is then reassaulted, the path of the sound bouncing erratically through the forest as if the cars themselves were out of control. It is not enough to see these trees, that they were big and old. I want to hear their silence. I want to simply listen to them drip.

Big-leaved maple

CHAPTER 25 FOUNDERS' GROVE

AS I got into my car to leave Founders' Grove in the waning twilight, another car pulled up. A man got out and I heard him ask someone, "How far is it to Founders' Tree?"

"Oh, about fifteen minutes," I heard the person call out the window as she drove away.

The man leaned back into his car, relaying this information to a man in the passenger seat and a woman in the back. They conferred, and the driver was on the verge of sliding back behind the wheel when I said, "Uh, excuse me. It's not fifteen minutes to Founders' Tree."

"How long is it?"

"It's one minute. It's right there," I said, pointing to the tree across the parking lot.

"One minute?" the man asked, to be certain.

"Yeah. It's *right* there."

"Can we look at it and get out of here by dark?"

"What do you mean 'out of here'? Do you mean this grove?"

"This grove? We mean these woods. Can we be out of these woods by dark?"

"Well, if you're just going to Founders' Tree, you can be out in *ten minutes*."

He leaned back into the car to relay this new information. They discussed it momentarily, and then the other people took off their seat belts and got out.

"I bet you'd come with us and show us around," he said. I found this man's bold friendliness refreshing. In a moment's time he had read my character, understood that I have a passion to share the natural world.

"Okay, I *will*," I said, accepting his challenge.

While I locked my car, they set off across the parking lot, down the short trail, and stood in a row in front of Founders' Tree.

"Could you snap a picture?" they asked as I caught up to them.

"Sure."

After the picture, they briefly read Founders' Tree's vital statistics and then turned to leave.

"You should see the Dyerville Giant," I said.

"Where's that?"

"Right down the path."

"How far?"

I felt like I was running a shuttle service.

"*Seven* minutes."

They looked at each other and conferred quietly in sketchy words, finally, reluctantly, agreeing to go.

"Where are you from?" I asked.

"I'm from Raleigh, North Carolina, and they're from Kansas City. We're from the *city*. We're not *used* to all this."

"I know what you mean. I moved to the heart of New York City for two years, and I had to get used to the country all over again when I left."

"How do these trees get so big like this?" the other man asked.

"It's all the rain. We get from thirty-five to sixty inches a year. This is a rain forest."

"This is a *rain forest?*"

"Yeah."

"Like a tropical rain forest?"

"No, it's a temperate rain forest."

"What does 'temperate' mean?"

I was touched by this man's curiosity, though he admitted to being so far out of his element.

"It means 'not tropical.' It's not in the tropics. We're in the temperate zone."

"This is a rain forest?" He continued to muse to himself, looking around. "This is *beautiful,*" he declared. Then he suddenly noticed evidence of one of the ancient fires that had burned through the grove. "This tree looks burned!"

"It's been burned many times in over a thousand years." I was beginning to sound like a docent.

We passed a sign, and the man stopped and tapped one of its lower corners, saying, "*That's* what I'm afraid of."

"I bent close to see what he feared, and there was the little bear, enclosed in a rectangle, that is the emblem for the California state parks.

"Oh, you don't need to worry about those." But I remembered vividly, after my two years of urban living in New York City, how utterly alien the natural world was for me when I left. Even though I had spent much of my childhood on the loose in the most wild stretches of the Mojave Desert, and camping was the one constant in my family's otherwise freewheeling existence, I had reached a point where a field at night, a forest, frogs, were unfamiliar and actually frightening. I couldn't imagine what this forest of nearly prehistoric trees looked like to these people in their rented car, pausing for a token encounter with the local environment, but I did know from experience that their point of view was vastly different from mine.

As we approached the Dyerville Giant, the man who had initiated my role as their guide said, "This is where those people in that university are putting those metal things in those trees." We were plumbing the depths of his awareness together. I was pushing them to wander in the woods, out to see a fallen giant, and in the process we had a captive moment to nudge cultures up against one another and compare realities.

"Spikes?"

"Yeah, spikes."

Then he said, "Yeah, and I've heard they're growing a lot more than trees out there . . ."

I hustled a little to catch up to his stereotypes, as he clarified, almost with a wink, " . . . a little weed?"

It made me sad to realize how the media misrepresents Earth First! In fact, Earth First! has vowed to maintain a nonviolent response to the violence of liquidation forestry. In Northern California, there has been one instance of tree-spiking in Mendocino County, for which Earth First! members adamantly deny responsibility. We reached the Dyerville Giant.

"Oh my God, it's so big," the woman said.

"Come look at the roots," I suggested, walking through the understory toward the base of the tree. "Those roots in the center began growing shortly after the birth of Christ."

"After the birth of *Christ?*" she asked, as if I might be talking about a different Christ than the one she knew of. The vertical mat of woody roots towered over us, reaching toward us as it had reached toward molecules in the soil over eons. The woman had left off her more passive approach to the trip and started saying, "Oh my God . . ." over and over.

We began walking back, and the shorter and quieter of the two men, the one who had not initiated contact, lingered, fixing his gaze on one particular tree.

"Look at that tree," he said.

No one was listening but me.

"Touch it," I said, impulsively.

"Touch it?"

"Yeah. Touch it." This was a bold request because it required that he backtrack in the opposite direction from his friends.

"All right, I will."

He walked away from us, directly toward the tree, through the twilight, in the drizzle. In the act of walking away, he gained an independent relationship with the forest. He was simply a man amid trees. He looked small, his black skin and the black bark of the rain-soaked tree the same color, as if he and the tree were both sculpted from the same wood. He walked until he was an arm's length from the tree, actually having to push into the damp huckleberry bushes. Then he reached out and touched the bark and looked up.

The other man and the woman had walked on a little ways, unaware of their lagging friend. They were seeing new things, but they had not changed realities.

The shorter man came scurrying to catch up with us.

"Look at that tree," he repeated.

"Which tree?" they asked him.

"That tree," and he pointed back at his tree.

"Yeah," they said, eyeing it up and down respectfully.

"No, *look* at it," he said.

"Yeah," they said again, with a little more emphasis.

"No, *look* at it."

At this point they looked at him instead. "Yes, we *see* the tree."

"I think it's as *big* as Founders' Tree," he said, as if physical size could validate his encounter with the tree. By then we were in sight of Founders' Tree, and we all looked back and forth between the two.

"No, I think Founders' Tree is a little larger," the woman said.

"I don't think so. I just feel like . . . *hugging* it. I'm going to be one of those tree-huggers."

His friends looked at him indulgently.

"I do, I just feel like . . . hugging it. . . ."

We had reached the parking lot. At that point we finally introduced ourselves, having to some degree shared realities. We exchanged business cards, and, as we shook hands, the taller man said solemnly, "This has been the highlight of our trip. Thank you."

CHAPTER 26 A MEMORY

WHEN I was a little girl about eight years old, my brother took me on my first backpacking trip. We went down Devil's Canyon in the San Gabriel Mountains behind Los Angeles. The trip was hot and grueling, one of those hikes where one goes downhill to get into camp and has a difficult climb up to get out. As I made my way back to the car, I was carrying just my mummy bag, my brother having taken the rest of my gear. A war surplus bag made of kapok, it was an object I hated. I hated being zipped into it. It gave me claustrophobia. And I hated pulling it up the trail, released from its ties, spilling this way and that out of my arms. It was one of those bags that was good enough for the youngest member of the family, and for some reason I never protested.

Though most of the trip seemed to involve getting back out of the canyon, my brother, with his long legs, yelling back at me in disgust to "just walk," I did have an unforgettable moment that seems to hang suspended all by itself in my dry childhood.

During our day in camp, my brother and I ventured farther down the canyon, my brother having heard that there were incense cedars growing near the creek. We explored, and finally we found them. To my eyes, they looked like redwoods. I felt that I had finally "arrived" somewhere that someone would actually want to be. They looked exotic, lush, like a promise fulfilled. And then, beneath them, we found a stand of tiger lilies.

If you did not grow up in dryness, I cannot tell you how sacred the plants and animals that rely on water can seem. To this day, frogs have remained nearly sacred objects for me, pouring their sound into my perpetually arid soul. Salamanders are even more exotic. And these tiger lilies, with their recurved petals, seemed like the flowers of paradise. We had hiked downhill, toward the L.A. Basin, yet it seemed that if we kept going we would arrive in Eden instead of Sierra Madre.

Now I live where tiger lilies are part of the local flora and redwoods grace our parks, and it sometimes surprises me to remember how my soul once craved what I so often take for granted. It is so easy to forget that redwoods do not grow everywhere. When I go away, particularly to the desert, I come back and for a moment see them: lush, mysterious, ancient, holy.

Western azalea

CHAPTER 27 THE FIDDLE

JUST when I was ready to call Earth First! activist Judi Bari, whom I had never met, to see if I could interview her, I learned that she had developed breast cancer and had declined treatment. A short time later I learned that she had died.

Judi was an ardent fighter for the redwoods. She moved to California from Maryland, where she had worked as a labor organizer. In Mendocino County she joined the fight against the logging of the old-growth rain forest, hooking up with Earth First! activists Darryl Cherney and Greg King, some of the earliest voices in the Headwaters issue. Judi had both a reverence for nature and a deep concern for people. It was not hard for her to see that the loggers and mill workers were in the same boat as the spotted owl, the marbled murrelet, the coho salmon, the southern torrent salamander, and all of the other species of the forest, not to mention the old-growth redwoods themselves. She preached job preservation through sustainable forestry and reforestation.

In light of history, it is ironic that Judi was highly influential in turning Earth First!'s tactics from midnight monkey-wrenching and tree-spiking, as championed by the movement's original founders, to Gandhian nonviolence. In 1990, college students and young activists were swarming to the North Coast for what was being billed as "Redwood Summer," a nonviolent gathering in defense of the forests similar to the civil rights demonstrations of the 1960s. The enormous appeal of this nonviolent approach began to alarm proponents of liquidation logging. Judi, Darryl, and Greg began receiving death threats.

At 11:55 A.M. on May 24, 1990, as Judi and Darryl drove through Oakland, California, a pipe bomb exploded beneath the driver's seat of Judi's car. She was driving. The explosion crushed Judi's tailbone; a spring from the car seat damaged her intestines; shrapnel embedded in her buttocks and pelvis, leaving her right leg permanently crippled. Much to Judi and Darryl's additional horror, the FBI immediately swooped in and accused *them* of bombing their own car.

It is interesting to note that prior to the bombing, according to David Harris's book, *The Last Stand*, the Eureka office of the FBI held a terrorist bomb school at College of the Redwoods, a junior college south of Eureka, for police and FBI agents in the Northern California area. One of the field exercises was to stage actual car bombings at a clearcut north of Eureka, using pipe bombs placed on the back seats of the cars. Three cars were blown up, one after the other, then investigated by the faculty for the edification of the students. FBI Special Agent Frank Doyle, the Terrorist Squad's bomb expert, was quick to underscore the fact that this type of explosion generally was the result of the careless transportation of bombs by the bombers themselves.

Within an hour of the bombing of Judi Bari's car, four FBI agents who were graduates of the Eureka bomb school appeared on the scene. Another agent who arrived was the bomb school's instructor, Special Agent Frank Doyle. He informed the Oakland police that despite the fact that its case was relatively intact, Judi's guitar had been placed over the bomb on the back seat, indicating that Judi knew of the bomb's presence in the car. The guitar case, supposedly laid over the bomb, survived the blast, yet directly *under* Judi's seat, the street was visible through a two-foot-diameter hole with edges curled outward, while at a nearby hospital, pieces of the seat itself were being surgically removed from Judi's buttocks and pelvis. Special Agent Doyle was quick to describe Judi and Darryl to reporters as terrorists and to remind them of Earth First!'s supposed connection to tree-spiking. While Judi lay in her hospital bed, she was arrested, and Darryl was put in jail.

Judi's own words in her book, *Timber Wars*, tell the story of the bombing in detail.

> *I knew it was a bomb the second it exploded. I felt it rip through me with a force more powerful and terrible than anything I could imagine. It blew right through my carseat, shattering my pelvis, crushing my lower backbone, and leaving me instantly paralyzed. I couldn't feel my legs, but desperate pain filled my body. I didn't know such pain existed. I could feel the life force draining from me, and I knew I was dying. I tried to think of my children's faces to find a reason to stay alive, but the pain was too great, and I couldn't picture them. I wanted to die. I begged the paramedics to put me out.*
>
> *I woke up in the hospital 12 hours later, groggy and confused from shock and morphine. My leg was in traction, tubes trailed from my body, and I was absolutely immobile. As my eyes gradually focused, I made out two figures standing over me. They were cops. Slowly I began to understand that they were trying to question me. "You are under arrest for posession of explosives," one of them said. And even in this devastated condition, my survival instincts kicked in. "I won't talk to you without a lawyer," I mumbled, and drifted back into unconsciousness.*

In the months following the bombing, the FBI lab came to a contradictory conclusion from Special Agent Doyle's. It found that the bomb was a motion device designed to go off as the car was driven, and that it had been placed *under* the driver's seat, attached to a piece of plywood. However, amazingly to some people and predictably to others, the case still remains "unsolved," and Darryl and Judi remain the only suspects.

After her death, I went to a service in Judi's honor in Willits, California. In the way I cross-reference birds, beaming images toward a central point in my imagination where a three-dimensional hologram slowly develops, Judi materialized for me—like a hummingbird gleaming with tiny iridescent green feathers, taking wing and hovering.

At one point, a talented young woman stood on the stage playing the fiddle, shifting her weight from foot to foot as she moved her bow with a robustness and vitality that for the moment put Judi up there before us playing. I sat on the grass entranced. How I would love to play the violin. It was the music of the soul circling over our heads, darting to our ears, effortlessly penetrating our minds and bringing the crowd to life.

There was Judi, beamed in, playing for all she was worth—though the fiddle on which she played was itself a ghost, confiscated by the FBI after the bombing and never returned. Music had been turned loose by theft and death to wing home as memory.

CHAPTER 28 ELLEN

I INTERVIEWED a tree-sitter yesterday, a young woman named Ellen Fred. Ellen is twenty-six years old, with a degree in Russian diplomacy from the University of Michigan. After college she went to Russia, and there she realized that her passion for diplomacy related not so much to people, but to forests. She moved to Siberia and worked on environmental issues in the Lake Baikal region. After returning to the United States, she became focused on the over-exploitation of American forests.

Listening to Ellen talk, I was speechless. Was this an Earth First!er? This was the beginning of my own slow climb out of the brainwashing that can affect a seemingly sympathetic person such as myself. Ellen is bright-eyed, thorough, succinct in her coverage of topics, cheerful with a tenacity that, as always when I am in the company of this sort of young person, has the effect of galvanizing *me*. She is America.

This is a book about America. When I talk with someone like Ellen, I hear the tune of "America the Beautiful" somewhere in the distance, a little beyond consciousness. It is the tune of hope and patriotism, humming itself, keeping itself alive, daring to wish us health and wisdom. It is not "The Star Spangled Banner." I do not hear, " . . . and the rockets' red glare, the bombs bursting in air, gave proof through the night that our flag was still there." I hear my own childhood voice at San Rafael Elementary School singing with idealism and confidence, "O beautiful for spacious skies, For amber waves of grain, For purple mountain majesties, above the fruited plain. America, America, God shed his grace on thee, and crown thy good with brotherhood from sea to shining sea." I find myself daring to adore America, because it still has the vitality to produce people like Ellen.

She spoke cheerfully, not quite matter-of-factly, about sitting for six days in an ancient redwood at the north edge of the Headwaters Grove. She was not pretending that it had been easy for her. Efforts to stop logging of ancient trees in a residual grove adjoining Headwaters had failed. Regardless of the feelings one might have regarding the sanctity of private property, it is riveting to imagine the courage of a tree-sitter. It makes one stop and ask, what have these young people seen that we have ignored?

Ellen spoke of her decision to do the tree sit. Arrangements took four days. With help she hiked in her gear on foot at night. I cannot imagine sitting seventy feet up in a redwood in the most serene of conditions. Ellen described how a climber scaled what was to become "her" tree, set the ropes, and then pulled her up into the tree through the darkness, using his own weight, as he simultaneously let himself down. She reached the platform where she was to stay for six days, and climbed into her sleeping bag to stay warm.

I asked her how big the platform was. She said it was an old door, suspended securely by ropes. She was okay as long as she was surrounded by the darkness of night, but as dawn began breaking, gradually illuminating the old-growth forest around her, she looked down and was terrified. She flattened herself to the trunk of the tree and gripped the fibrous bark like a baby monkey on its mother's back. It was too late to leave. The loggers would soon be breaking the silence of dawn as they arrived in their pickups, warmed up their chain saws, and the giant Caterpillar tractors began loading trucks with logs.

Ellen did the tree sit with one other person, a young man named Chris who sat across the logging road in a tree that was slated to be cut down. A traverse line linked their two trees, and each day, the logging trucks passed beneath this line, over which Ellen and Chris were able to travel at night.

I asked Ellen if she became very attached to her tree, and she grew solemn, slowed her speech and looked at me.

"My tree was wonderful. I hung there by her side, watching the logging coming closer, and I understood logging from the tree's point of view. Like the trees, I couldn't move. Behind Chris there was increasing daylight through the canopy, until finally there was only his tree and the one beside it still standing on that side of the road. The loggers told him to come down. Chris wouldn't do it."

She looked at me hesitantly. "At one point we were yelling at the loggers, trying to get them to stop, and this one man looked up at me and yelled, 'Fuck you, Gaia.' I just stopped and cried. Finally a logger began sawing the tree right beside Chris. When it was almost cut through, the workers got in their trucks, leaving it teetering as they drove home for the night. A logger stayed behind and begged Chris to come down so he could cut the swaying tree, but Chris refused.

Ellen, clear-eyed, precise, looked at me and said, "We were ready to die." Of this I had no doubt. She went on, "The logger began sawing the tree, working to aim its fall as far from Chris's tree as possible. It fell, barely missing its last neighbor, which then stood by the side of the road all alone, with Chris dangling from its side."

Is Chris's tree still standing?

"No, they're both gone." She was the elder in the conversation, informing me about the reality of industrial logging.

"They're *both* gone?"

"Yeah, they've logged the stand that I was in too."

"How did you decide to come down?"

"They simply went past us. Our trees were left all alone surrounded by a clearcut. There was no point in staying anymore. They had won. So we climbed down with our gear after dark and left."

I was amazed that Ellen has experienced a nagging sense of failure over this experience, a feeling that she let her tree down, didn't uphold her responsibility. This is a private feeling, not to be consoled or diminished by anyone else's expressions of admiration or appreciation for what she did. It lives between her and the tree.

CHAPTER 29 CLEAR-CUTTING

PERHAPS men cannot help clear-cutting. The technology is too new to have been tempered by natural selection. And the drives that propel men—to provide for family, to be the biggest, to conquer the most—are perhaps too old to be easily corralled. Technology has put powerful tools in men's hands. A Fish and Wildlife biologist overheard loggers, unaccustomed to being asked by their superiors to clear-cut, stand back and survey a wasteland of their own creation.

"Wow, can you believe we did that?"

I wrote the preceding paragraph in a tolerant mood, imagining a young man for whom simply having a job and being responsible to his wife and children represents an enormous step in personal growth which is not to be minimized.

But a friend who read this paragraph wrote on the bottom of the page:

Anyone can stop wholesale slaughter whether it's bunnies or harp seal pups or our last surviving ancient forests. When it's human blood that's spilled, we put the perpetrators away. We see a man beat his dog and he goes to jail. It isn't because we're against letting dogs go extinct, it's because we can't bear to witness it. We just don't apply the same standards to the wild and mysterious as we do to the domestic and known—unless we are sickened by witnessing the violence.

Clear-cutting is an act of violence that affects trees, rivers, air, water, earth, and every person, owl, toad, or human who lives there. It is violence excused by economic imperative. I can't say, "I needed the cash so I harvested the people at the 7-Eleven and took the cash." Why should Hurwitz?

I think she's right and I'm wrong, but I have still known a lot of fine young men caught in the middle.

CHAPTER 30 FISHER GATE

I AM faced with three more hikes. The one Doug is proposing now takes us past Fisher Gate in Carlotta, and through Yager Camp, one of PL's principal bases of operation, where logs are stacked out of sight before they are taken to the millpond in Scotia, and logging trucks are repaired in gigantic open-bay garages. He is hoping it will not only be raining, as it was on our first trip, a steady, soaking, miserable rain, but he is hoping it will be pouring so that the fifty-foot waterfall on Allen Creek, which few people, including PL employees, have ever seen or even know exists, will be crashing with water. The waterfall is in the middle of Allen Creek Grove, where Doug just happened to come upon it with a friend, never having even heard of its existence.

Doug has attempted to reassure me, at least from his point of view, of the relative ease of trespassing straight through the middle of Yager Camp.

"Is there a guard?" I asked.

"Yeah, sometimes. And there *are* dogs there. You can hear them barking, but I've never had them come after me."

"Dogs?"

"Yeah, but they're chained, I think. But they do sound big." People of my persuasion do not do things like this. We choose other demonstrations of courage than all-out athletic feats or exposure to physical danger.

There is additional tension associated in my mind with Yager Camp, because it lies just on the other side of Fisher Gate. For me, Fisher Gate is a local moral black hole that sucks good karma into the bowels of the earth. It was the scene of the November 15, 1996, rally that David and I attended shortly after I agreed to write this book. A description of that rally will not only give you a better sense of the rather unique political climate of our county, it will set the scene for our night trip past Fisher Gate and through Yager Camp.

At the time of the November 15 rally, I sensed that there was some degree of residual petulance among law enforcement officers following the success of the September 15 rally just two months before, which not only spawned Taxpayers for Headwaters but attracted seven thousand demonstrators, making it the largest forest demonstration in the history of the United States. At that earlier rally, over one thousand people, including musician Bonnie Raitt, stepped over a symbolic painted line that represented the PL boundary, and were ceremoniously handcuffed and escorted one by one to be booked at a nearby desk. National news trucks lined the road, beaming to satellites above.

September 15 marks the official end of the marbled murrelet breeding season and the renewal of logging in the hills behind Fisher Gate. This was the third September 15 rally staged at Fisher Gate, each one doubling and tripling the numbers of the last. With so many people around the world watching this massive nonviolent rally, the police could do little more than passively stand by or tediously process the astonishing number of people waiting patiently in line to have their wrists bound with plastic handcuffs.

After I attended this gigantic rally and subsequently read on the plaza in Arcata, David and I went to the rally in November to sell the murrelet T-shirts I had had printed. The minute we arrived, however, we were instantly on edge. We both had a distinct feeling that the rally's organizers had committed a tactical error. Fisher Gate is on a lonely, dead-end country road in a redneck town where nearly every yard boasts a sign on a picket that reads, "This family supports the timber industry." On the front of the small store in the middle of town, there is a sign that reads, "EPIC and Earth First! not welcome." Because the rally was on a Friday, not a Sunday, there were no satellite news trucks; Bonnie Raitt and Bob Weir were not there singing; the small crowd was made up mostly of young people.

"So, why was the rally here at all?" you might ask. Because this is one of the principal gates through which PL's logs are brought out of the forest. I was momentarily comforted when I was handed a print-out that clearly stated the details of the rally that had been worked out at a meeting between organizers and the police. I read out loud the succinct directions on how to avoid arrest or how be arrested uneventfully. I was relieved to see that avoiding arrest was not hard. One simply had to sit down by the side of the road and look uninvolved.

But we continued to watch events with growing uneasiness. Something was wrong despite the handout and the cheerful demeanor of the crowd.

Many people held cardboard cutouts of coho salmon that swam on sticks above the crowd, which gently swayed with the music around a flatbed truck serving as a stage on the shoulder of the two-lane road. These fish contributed some levity: I heard someone refer to the fact that the rally was a "swim-in." Yet, police cars kept ominously parting the waters, cruising back and forth through the crowd, which fluidly refilled after each passage and kept swaying. Perhaps an hour into the rally, Naomi Steinberg, a local Jewish leader, spoke compassionately about the logging families of Scotia, where her husband had taught school for many years. Darryl Cherney began singing and the fish bobbed rhythmically. A few more police cars parted the salmon, and then we saw the only news media car in the area suddenly speed away with its windows rolled up, its two occupants looking neither right nor left at the crowd. At that moment, feeling some balance had been tipped beyond our tolerance, David and I began hastily folding up our shirts and our card table as we both said, "Let's get out of here." Later, demonstrators I interviewed reported having overheard the police tell the media to leave.

The sound of the music began to fade, and we felt peaceful again as we walked down the narrow country road lined on each side with Himalaya berries and barbed wire, with open fields beyond. We reached our car, loaded up our table and our box of shirts, and drove away, shaking off what we imagined was "just us," just our own hypersensitive, chickenhearted feelings of "bad vibes." The next day, however, through the grapevine, *not* reported by the press, a gruesome story began to coalesce.

At the end of the road, near Fisher Gate, a large black tarp had been suspended, providing a screen. No one knew that behind that tarp, the police were readying chain-gang-style chains with handcuffs. Just moments after David and I left, the police, on cue, unprovoked, lined up and abruptly advanced at the same time that a hasty announcement was made that the rally had been declared an "illegal gathering." Most people were still listening to the music and didn't hear the announcement, while others were suddenly grabbed and handcuffed to the long chain. Photographs show relaxed body language in most of the people, juxtaposed with horrified grimaces of the individuals being suddenly seized. Bystanders with cameras were targeted and their cameras confiscated "as evidence."

One girl reported to me that, following the directions on the handout, she sat down on the grassy shoulder of the road, indicating that she did not choose to be arrested. She was a new student at Humboldt State who attended the rally "to learn." She had heard about Headwaters, and she simply felt obligated to know more about the issue. She was pulled up from the grass, however, and handcuffed to one of the long chains that the police were dragging over the pavement. When her chain was "full," they were led behind the black tarp, out of sight of the rest of the crowd, and their chain secured. In a handwritten account, with lovely, well-schooled penmanship and impeccable spelling, the girl described how a little after three o'clock in the afternoon she told a policeman that she needed to go to the bathroom. He told her she couldn't. She asked again an hour later and was told the same thing. She said that the wrist band was cutting into her arm, and she asked a policeman to loosen it. He said, simply, "Tough." She sat on the ground with the others on that chain from three o'clock in the afternoon until midnight, when she and her chainful of fellow captives were transported by truck to Eureka to be booked. When she asked, in the Eureka jail, if she could go to the bathroom, she was shown a toilet in view of the booking desk. She spent the night in jail on charges of "resisting arrest."

I was told that other demonstrators who scurried away from the rally shortly after we did were pursued by police in cars. Several other people told me that after they were already driving away, they were stopped, their car doors opened, individuals pulled out and arrested, and more cameras confiscated. Demonstrators who had been arrested nonviolently at the rally at Fisher Gate two months before, this time were seized as officers leaning into their cars pointed fingers and called out, "Repeat . . . repeat . . ."

I feel that I must offer the disclaimer that I know that there are many fine policemen and sheriff's deputies in Humboldt County, just as there are many fine employees of the timber industry, and many fine people trying to understand what is going on behind the locked gates of MAXXAM land. Even though the event was not even reported in the media, numerous people who were at the November rally have submitted affidavits to the American Civil Liberties Union reporting what they experienced. To everyone I interviewed, it was obvious that the police attack was premeditated, designed to intimidate those people who have the courage to lawfully demonstrate on behalf of our forests and those who, like the one girl I interviewed, choose simply to inquire.

There are those who shrug their shoulders and call Humboldt County "Hurwitz County," and those who will not accept that designation without a fight.

Douglas squirrels

CHAPTER 31 ALLEN CREEK GROVE

AT midnight, at Fisher Gate, standing before the No Trespassing sign, I panicked.

"Doug, I don't have the courage for this."

Somehow I thought that was all I would have to say, and Doug would say, "Oh, I understand. Of course you don't. Come pick me up in twenty-four hours."

But as soon as he matter-of-factly said, "Too late now," and began climbing over the gate, I realized it would be too hard to argue, there in the moonlight, and I climbed over too.

David came with us, because I had said to Doug earlier, "I can't go. There's nowhere for David to stay." Suzanne could stay at a friend's house, but I couldn't make arrangements for David.

"Bring him," was Doug's carefree response, as if we were going for a picnic in Humboldt Redwoods State Park. The Owl Creek hike had been lots of fun for David, I reasoned, and I had had misgivings about that trip. He is learning an enormous amount about forestry, land use, politics, biology, human nature, his own endurance, and, if I look at the situation in perspective, his own mother. But I dreaded "the dogs," real or imaginary.

"They've taken the vocal chords out of them so you don't hear them till the last minute," Doug teased me. "No, that's true," he said, now thinking about dogs, not Fisher Gate. "I read about that. Some people in some city kept having their guard dogs shot because the prowlers would hear the dogs coming and shoot them, so they had their vocal chords removed. You'll just hear the clickety-clickety of their little claws before they bite," he said as he scampered his hands like paws in a momentary imitation of an advancing Doberman. Children love Doug because he never misses an opportunity to animate a story or add sound effects.

In hindsight, it might have been better if I had not brought David, but I guess this is true of many adventures. If we knew what was going to happen, we might not ever go. On the other hand, another friend of mine simply uses the term "pub stories." When he was stuck in a life raft for three days out at sea, it was a "pub story." Whether we knew it or not, David, Doug, and I were embarking on an adventure that I could imagine David recounting in ten years as he tried to describe his childhood. We walked down the quiet road, and I felt overwhelming dread. David's hand felt small in mine as we made steady progress toward the bright lights of Yager Camp.

On the edge of the camp, we came to the welcoming cover of towering log decks and ducked into their protection, making our way down narrow aisles between eight-foot-diameter logs. As I watched David pick his way ahead of me among the rain-filled ruts of Caterpillar tracks, I got one of those blasts of love that one gets at certain moments as a parent. I recalled how he had earlier begged me not to trespass. At the time, that seemed not only like a reasonable request, but I was of the same opinion. The last thing I thought I'd be doing was sneaking with him through the corpses of ancient trees. It was like being among whales or in a concentration camp; these are the only two analogies with appropriate magnitude.

We paused at the far edge of the log deck surveying the scene. The open ground was disturbed by the repeated passage of Caterpillars and littered with the white tags that are normally attached to the butt ends of the logs. These tags, scattered, discarded, looked like the name tags of prisoners who have been "disappeared."

Finally there was no choice but to rejoin the road and head straight through the middle of the camp. We paused on the edge of the night, eyeing buildings and floodlights, watching for movement and dogs. Doug was concerned because, while he had crossed through Yager Camp countless times with other people, with reporters even, he had not done so for a year and a half. He could see there had been a lot of changes: more concertina wire, more chain link, more lights. I had run through the worst-case scenario in my mind, a five-hundred-dollar fine and a visit to jail. I imagined that I would be open about my mission, explaining to the security guard that I was an author attempting to write objectively about the Headwaters issue. I figured I would interview our captor and, if I could, engage him in a discussion about his job and the changes he had observed at PL since he began working there. I would make it clear that I was not there to "monkey-wrench" equipment or whatever PL's worst fears might be. But this all collapsed when I imagined facing a Doberman, trying to put myself between the dog and my son. "How do you like your job?" just would not cut it.

Wordlessly, Doug set off, testing the waters, leaving us behind on the edge of the shadows. It was about midnight on a Sunday. Rain that had been falling for two days had let up. The camp appeared quiet. I didn't want to let Doug get too far ahead, so after studying his progress for a minute or two, David and I shoved into the light like possessed skaters, holding hands, almost gliding in our effort to make no abrupt moves. Our eyes cast left and right, scrutinizing doorways, shadows, vehicles, and if our ears could move like horses', they too would have tracked our fears as we listened for that dreaded growl or bark or that "clickety-click." But apparently we were alone.

We reentered the night after about a quarter mile of stone-faced walking, and then we came alongside Yager Creek. Because of its name, I had underestimated its presence. It is a river. A creek is more private. A river has a public quality and commands a different respect. Because of the rain, Yager Creek was silt-laden and bounding wide between its banks. I had never dreamed that there was another river of such size and presence near my home in addition to the Van Duzen, the Eel, and the Mad. It is a significant moment when one meets a river for the first time, and we stood, absorbed by its power and character.

We walked along Yager Creek for a couple of miles. After the trip to Headwaters, I am pleasantly surprised by any land that is not stripped bare. In the buffer zone that remains along the lower reaches of the river, the second-growth trees are tall, and there are still small stands of old-growth as well. As if in confirmation of the river's beauty, the full moon made an appearance and danced along through the treetops. Doug has repeatedly stated that the Headwaters "Deal" is incomplete without Yager and Lawrence Creeks and their watersheds. Increasingly I understand this point of view. Perhaps the timber companies only think in terms of board feet, but the creatures of a forest rely on whole ecosystems, within which they are uniquely interdependent. The rivers with their wild salmon are a critical part of this whole.

The Headwaters Deal was brokered by U. S. Senator Diane Feinstein of California in September, 1996, in an effort to reach an agreement to protect Headwaters. "The Deal," as it is known, will protect through public ownership 3,117 acres of virgin old-growth forest, 665 acres of residual old-growth, and 3,693 acres of second-growth and cut-over lands. This includes the Headwaters Grove, Elk Head Springs Grove, and the South Fork of the Elk River for a total cost to the federal and California state governments of $380 million. Many people, including Doug, who served as an expert witness during negotiations in Washington, D.C., feel that this is not enough. They see it as only a token preservation of an outstanding area. They believe that 60,000 acres is the minimum that will protect viable ecosystems, not just fragmentary examples of forest, or "tree museums."

Sometime around three o'clock in the morning we turned uphill out of the river gorge and climbed toward Allen Creek Grove. David is a singularly uncomplaining hiker. Ever since his first backpacking trip at the age of four, when he hiked a total of fourteen miles to Ward Lake in the Trinity Alps, he has thrived on covering ground. Self-employment has its benefits. A child experiences the interests of the parent more personally, I think. Parent and child pull together as a team. David and I pulled the cause of Headwaters up the slope like oxen, until at five o'clock A.M. we reached Allen Creek Grove, laid out our sleeping bags, and went to sleep.

At seven o'clock I awoke to rain falling on my face. I looked over and saw David's face wet with drops. Half awake, I hastily put up my tent and got David inside without him even waking. Doug stayed out under his tarp and at ten he woke us up saying it was time to go see the waterfall. I looked at David sleeping and my heart ached.

"Doug, we have until five o'clock tonight before we can even leave here. Why don't we go this afternoon?"

"Because right now it isn't pissin' down with rain." That was clear.

I woke David with a promise that he could sleep later, and we unzipped the tent and found ourselves surrounded by beautiful ancient trees, pale green with lichen. I just wanted to sit and absorb all of the intricacy of form around us, but we left the tent and our packs, and continued on upslope toward the top of Allen Creek Grove. By daylight I could see how cut-over the landscape is, with the exception of the immediate river valley along Yager Creek.

As I moved up the steep, seemingly endless hill, my desire for conversation dampened by lack of sleep, my mind sank its teeth into the devastation around me. All I could think was that it is simply "bad business" to destroy our own forests in 150 years, selling raw logs overseas. It is simply "bad business" to use trees for paper. Sixty percent of the trees that are cut down go for paper. More appropriate fibers exist, such as kenaf and hemp. It is simply "bad business" to print large daily newspapers that few people thoroughly read. It is "bad business" to not slowly mete out our old-growth redwood and sell it at premium prices, maintaining this ecosystem in viable condition. No rancher who liquidated his own breeding stock, no matter how much money he was netting in the short run, could be considered a good businessman. He would be considered suicidal, or at least frivolous. With over 90 percent of all types of ancient forest in the United States gone, maybe it's bad business to use *any* old-growth trees. It makes more sense to rehabilitate our second-growth forests and actively practice sustainable forestry. "Not one more ancient tree! Not one more ancient tree!" I heard the voices of young people at the demonstrations like a mantra.

Why do people admire a businessman who buys a forest in prime condition and runs it into the ground within a few short years, when those same people would probably view him with contempt if he did the same thing to a car? Hurwitz is like someone who has bought a fancy car that is so out of his price range that he cannot even afford tune-ups or oil changes. In order to service his enormous debt, he must clear-cut. If he clear-cuts, he must spray herbicides to control the brush. If he clear-cuts and sprays herbicides, erosion and poisons become the business of all of us downstream . . . and we are all downstream.

Finally we reached the upper edge of Allen Creek Grove. I was shocked to find that even though it is one of the six virgin old-growth groves left in the 60,000-acre Headwaters complex that environmentalists are trying to save, it has recently been crisscrossed with skid trails for the extraction of salvage logs. Doug got that same look that I saw when we were at Owl Creek Grove. He was swallowing hard, accepting change as we traversed steep slopes trying to find the elusive waterfall. Plastic ribbons fluttered everywhere, spotted ones, striped ones, red ones, yellow ones, white ones, blue ones. Some read, "Timber Harvest Plan Boundary." Others marked marbled murrelet survey stations or spotted owl calling stations. Mischief was everywhere throughout the grove. Finally I stopped searching for the waterfall. After several hours of arduous bushwhacking up and down hills, I was getting cranky.

"When you find the waterfall, yell to us or come back and get us." David and I sat down, and I began collecting a pile of lichen-covered bark and twigs with lungwort from out of the canopy that I thought I'd send to my mother in a little box.

Why do I feel a need to prove to my mother what is being lost? What can she do? Yet I feel a need for her to say to me, "What a tragedy. What a senseless waste. I'm glad you are writing this book." In a matriarchy of spirit, I need her blessing based on firsthand proof. I cannot bring her up to Allen Creek Grove, but its quality is expressed in its tiniest life forms as much as in its trees. Lungworts, which do not grow until a tree is hundreds of years old, which fix nitrogen from the air and drop, decomposing, into the soil, will look unearthly coming out of their tiny package on my mother's lap in Pasadena. Threads of lichen a foot long bespeak only time. My mother will put these twigs and pieces of bark on one of her end tables in her living room, and she will study them and show them to her friends. My child heart needs the case of Headwaters tried in her living room. It needs an acquittal, apart from what actually happens to the forest.

Finally we gave up on the waterfall. We could have found it if we'd had more time. Doug later went back to the grove and searched for it and found it just downstream from where we had been. I never had the heart to go back through Yager Camp to see it, which I regret. It is fifty feet tall, as tall as a five-story building, and unique perhaps in the entire world. I have asked people if they know of any other falls of this size entirely surrounded by ancient redwoods, and no one has. Doug has described the lush vegetation kept perpetually damp by the spray, the vine maples, the five-finger ferns, moss and lichens draping off the trees, the giant trilliums. For him it is a profound example of just how many undiscovered treasures may still exist in the Headwaters groves. But my knee had begun to hurt when I walked downhill, and I was concerned about how I was going to make it down the steep road we had taken the night before. So we returned to the tent, took it down after a brief rest, and packed up. The valley was silent. The loggers had gone home, but the logging road still seemed swept by speed. It looks like a racetrack, packed so hard by the weight of the logs that one's feet and joints ached, yet thinly layered over the top with a fine, chalky beige mud, utterly smooth except for barely incised tire tracks. Because of the poison oak that lines the road, when we needed to rest, there was nowhere to put our packs but on the road, so the mud began to spread to our clothes.

We reached Yager Camp before dark and sat in the bushes beside the road. After dark we approached the lights, and we could hear voices. Headlights moved. My heart sank. A radio played country-western music. *What* was I doing? Why was my beautiful nine-year-old son on the wrong side of Yager Camp? Couldn't I possibly think of anything else to be doing with my life? I comforted myself that at least Doug would realize that sneaking through was out of the question. This was a relief. The high-pitched beeping of large equipment moving in reverse reached my ears. Good. I didn't know how we were going to get home, but it was clear we were not going back through the lights.

Doug had said that it was difficult, but possible, to skirt Yager Camp by bushwhacking through the huge thicket that lay silhouetted between the road and the river. I heard frogs. Funny how one's values change in an emergency. Though I normally love frogs, their sound disheartened me. It meant standing water. I felt as if we were looking for a way straight into an immense swamp. We edged along the darkness, all the while getting closer to the lights.

A few days earlier, on the phone, a friend named Jesse Noell had spoken in his slow, considered voice about salmon, which had led, over the course of an hour-long conversation, into a hypnotic description of Gaia. He described a force that is always at work to keep life on this planet. As I listened I was sitting outside, staring at a hyacinth that had pushed its way up through a coiled garden hose on my lawn and bloomed as if nothing were unusual. Gaia. I heard the word in a new context, through Jesse's belief, and it took on new significance. Though it had been available to me as a concept for years, I was suddenly given a new god who is female. I had been looking for this.

As I surveyed the dense thicket, I began a silent chant: "Gaia, wrap us in your golden light. We are doing the right thing. Gaia, wrap us in your golden light. We are doing the right thing. . . ." The distant sound of country music, issuing from within the harsh glow of mercury vapor lights, made me feel even more like an outsider. Our feet started moving and step by step the thicket parted. Shadows revealed openings, and we picked up a Cat trail that wound its tread like a friend. "Here, over here," it beckoned. "Gaia, wrap us in your golden light. We are doing the right thing." As David and I skirted the familiar scene of our passage the night before, the trail would falter. . . . "Gaia, wrap . . ." and then resume. Poison oak became a friend, bending over us, hiding us. Doug came and went from view, slipping along ahead of us, his clothes and pack as dark as ours. Periodically, David checked to make sure his light-colored T-shirt tail had not slipped out from beneath my black fleece that he wore. "Gaia, wrap us in your golden light. We are doing the right thing." We looked into the open mouth of a giant repair garage where men were working on log trucks and neatly rhymed songs about broken hearts kept pace with the work. We paused on the bank of a small creek bed, eyeing the garage, and then dropped down one by one, crossing over round, tumbled stones, and scrambling up the other side into the trees. "Gaia, wrap us with your golden light. We are doing the right thing."

We stopped and surveyed a large clearing that contained some strange little trees—a future "demonstration forest," intended to show visitors to Yager Camp what kind of order exists farther upslope. In actual fact, given what we had seen, they were funny, incongruous, hard to take seriously. What exactly were they? Three-foot-tall trees, all the same shape, evenly spaced in rows. We bobbed in and out of regularly spaced depressions in the soil made by the bulldozer that had cleared the field. Despite their regularity, the depressions were a little off-stride, which made walking maddening. It was as if we were comic-book characters, crossing through a comic forest, a mockery, yet, perhaps someday, to someone newly arrived from the city, a mini-landscape that would look like a model for the future.

Once on the other side of the trees, we joined the road and found Fisher Gate standing open. No need to climb out. One last favor. Thank you, Gaia. Goodbye, invisibility. Hello, America, imperfect as you may be.

CHAPTER 32 DRAWING OWLS

ONE DAY I was teaching children in a school classroom how to draw owls. Children, after the age of about seven or eight, have a desire to create images that look like "reality." They know what this is and so do I, so we ignore all of those people who would ask them to be satisfied with simply "expressing themselves," and get down to the brass tacks of making drawings that look like the real thing.

I was leafing through my "owl folder" as I spoke to the children and came upon the riveting stare of a great horned owl. I spontaneously had an idea and found a picture of a spotted owl as well. I wanted to make real that word "owl," so abused on bumper stickers.

SAVE A LOGGER, EAT AN OWL.

WHEN YOU RUN OUT OF PAPER, WIPE YOUR ASS WITH A SPOTTED OWL.

When I see these slogans, what hurts me most is to imagine a child silently staring out the window of a car, absorbing these words with his or her other fledgling reading skills.

Wipe your ass with a spotted owl? What does that mean in a child's mind? How many owls has that child ever seen or even heard? I have taken my children owling at midnight with biologists who are so owl-like themselves, hooting with such mysterious authenticity, that it almost seems superfluous to hear a real owl reply. But they do, far out in some tree in the darkness. The spotted owl barks; the great horned owl hoots in its distinctive cadence, "Long, short, short," and then a varying number of hoots after the two short.

First I held up the picture of the great horned owl and walked slowly around the room. This was a lesson in "eyes" and the effect of pigmentation. The great horned owl has compelling eyes, yellow with a sharp black pupils, that look like their owner could eat a seven-year-old child. Indeed its prey list is long, including large animals such as house cats, pheasants, and porcupines. I have even worried about my small dog being out at night. One of

Northern spotted owls

the tragedies of clearcuts like those surrounding Headwaters is that the spotted owl and other smaller birds cannot cross the bare ground without serious danger of being attacked and eaten by great horned owls. Though the great horned owl is only five inches longer than the spotted owl, twenty-two inches compared to seventeen, it buries its large talons in the back of its prey and squeezes, usually killing instantly.

Then I held up the photograph of the spotted owl. It is a very different image. Its eyes are dark brown and doelike.

The children all exclaimed with sweet ahs and oohs. It is a happy chance of nature that this much-maligned bird has such endearing eyes. Of course, if you were a mouse, you would not think it was darling at all. Studying the eyes, however, was a way to put the owl itself before the children, so that when they read their next bumper sticker, they would have an image of something real, and not just adults' political symbols.

We drew just the eyes, putting in the gleams that children love as if I taught them magic. Then the children went ahead and extended the owl out around the eyes, feathering the successive rings that radiate like ripples across the discs that surround the eyes.

I told them a story as they drew, about going with David and Suzanne to see spotted owls with a biologist from Simpson Timber Company. The trip had been arranged by a friend who frequently hires me to do drawings for our local Audubon Society chapter. He wanted to create a T-shirt with spotted owls performing a mouse exchange. During the breeding season, while the female is on the nest, the male catches a mouse and calls to the female, who leaves the nest and perches on a branch. The male then flies to her, edges closer, reaches his head forward, and passes her the mouse. She flies off with it, bites off the head and eats it, and then eats the rest of the mouse or takes it to her young. As it turned out, the design never got used. The spotted owl, here in Humboldt County, is just too controversial, even for an organization founded on the love of birds. But my children and I got to sit in the forest and have owls repeatedly fly to our feet.

The spotted owl is one of nature's most "trainable" birds. Once a pair of owls has been called down to a certain spot and fed there, they remain eager to return at a moment's notice. David rode in the biologist's pickup truck and Suzanne and I followed in our car, being admitted through the locked gate, onto Simpson land. We switchbacked up the hill, the biologist parking at a couple of owl-calling stations along the way, until his hoots were returned from somewhere out in the woods.

"Hooo hoo-hoo hoooah."

With that reply, he immediately set to work loading up a length of perforated plastic pipe, capped at one end and threaded on the other, with white mice from a cage in his truck. He screwed on the pipe's lid, and we climbed down the hillside off the shoulder of the road, carrying our little capsule of life. He called again, "Hooo hoo-hoo hoooah," as he simultaneously stooped and picked up a five-foot-long stick.

I was shocked to see one owl fly right in and land on a branch directly above us, and then another. How could it be this easy? It was past the breeding season, so we were not likely to see a mouse exchange, but we watched, completely absorbed, as the biologist gently pulled a mouse from the canister and set it out on one end of the stick. The extremely white mouse tamely explored the stick, wandering innocently farther and farther out toward the tip. The biologist then lifted and extended his arm, putting the mouse far out into space. Both owls were leaning forward from their perches. The moment the end of the stick stopped rising and only the mouse itself was moving, one of the owls flew in, deftly plucked up the mouse, ascended to a branch, and ate its head off.

"Want to hold the stick?" the biologist asked Suzanne, rebaiting it.

"Uh, I guess."

She raised the next little mouse owlward, and just as fast, the other owl swooped in.

David took his turn holding the stick, and then we sat down, and a mouse was put directly on the forest floor in front of us. It scurried right and left, momentarily erasing the color of the leaves as it made sense of its rich new surroundings. Both owls watched fixedly, gauging our positions and the mouse's, and then one silently swooped in, lightly striking David's knee with its wing, making holy contact between owl and boy, before lifting off again with its prey.

Just because the spotted owl is strangely "cooperative" and "friendly," that does not mean that it is less in need of adequate habitat in the wild. I felt ashamed, watching the owls. Despite our use and misuse of them as symbols and tools, they are simply owls, living in innocence of our plans and manipulations. They want only a limited territory and prey.

I thought of the man who came to a writing class that I recently taught as a guest artist at Humboldt State University. He was older than the rest of the students. The class was about nature writing, and all of the students won my heart with their passion to express what they were finding was close to inexpressible. Words were not working for them. This one man pointedly asked how to weave fact and spirit. I could feel his urgency, and after the class, as he came up to shake my hand and thank me, I inquired, "What do you do for a living?"

"I work for Fish and Game."

I asked him his name and it turned out that he was one of the men it had been suggested I contact for an agency point of view. I could not have been more surprised if the owl itself had sat in the back row of the class.

"I write reports regarding PL land."

The miles that Doug and I have hiked on foot, he drives in a pickup, surveying the same destruction, one lone man trying to hold a balance.

"I want to learn how to put more feeling in my reports. We are expected to simply write dry science and there is so much more that needs to be expressed."

We are in the front lines of a battle. One never knows who will be alongside in the trenches. I would like to go back to that class and visit with the students more. They are mostly wildlife majors, among them, future Fish and Wildlife biologists, game wardens, perhaps a President of the United States.

Perhaps a future President drew owls that day with me in this elementary school classroom. Through art, the soul has the power to reach its filaments into the world of science, and restore balance. Art, just like man, without nature, runs the danger of isolating itself from life itself, just as science, without soul, does not remember its own reason to be. A President who draws owls . . . I think so. An agent for the California Department of Fish and Game who aspires to remind his superiors of the soul of the land he has traversed . . . I think so. Gaia, reaching her fingers to massage even the minds that connive to acquire more than they need . . . I think so.

Crimson columbine

CHAPTER 33 MEETINGS

SINCE I began this book in November, I have become friends with a fascinating range of people, not just activists, but also the rural people in the river valleys whose headwaters are on MAXXAM land. I began to focus on the various watersheds when a friend who lives on Cummings Creek told me about the destruction of her road and the loss of integrity of her creek caused by logging. A meeting had been called at the local elementary school in Carlotta, so I went and listened to confused people blaming one another for flooded basements and eroding backyards, and a road that kept washing out with each rain.

It was quickly established that the road, which was originally built for hauling timber, had been improperly laid down within the gorge of the creek rather than up on the mountainside. But this didn't account for the sudden, repeated washouts. That road had withstood much more severe weather than we had been experiencing without caving in or sending debris torrents into people's backyards.

PL executive Tom Herman began patiently explaining how the channel of the creek was clogged out on the floodplain, as if by some sort of moral failure rather than debris.

My head imperceptibly jerked like a dog with ear-mites. Downstream? Floodplain? The residents were well aware that that the mountainsides above them were being savagely clear-cut. These clearcuts were crisscrossed with skid trails, zigging and zagging, tracing a desperate pursuit of trees. After the logging is finished and the trucks have gone, each skid trail retains the potential to channel water and continue cutting into the earth itself, scouring an ever-straightening path until the stuff of the forest becomes the enemy of the people downstream. I listened to the anguish of a once peaceful community blaming itself for not having kept the ditches of its floodplain clear, and finally raised my hand to make a suggestion, "If I lived here, since water runs downhill, I would start at the top of the watershed to begin looking for the sources of the problem. I would try to project future sources of runoff over the next ten years."

This struck a chord and some people briefly ran with it, until Tom Herman artfully drew attention back down to the floodplain.

When I saw how easily attention was diverted from what seemed to be the obvious problem, it occurred to me that the people of Cummings Creek needed to see their own suffering mirrored in the lives of their counterparts positioned along other creeks just ridges away. Suffering could be turned into action once it was shared, so I asked around and found out that, coin-didentally, the community of Elk River was about to have its first watershed meeting the next week. I called Ralph and Nona Kraus, who live along the Elk River, and they invited me to attend the evening meeting.

When I got up to leave Suzanne and David in the middle of dinner, announcing that I had to go to a meeting, they looked at me sadly, stoically. Yes, they would clean up the dishes and leave my plate of food for me to finish later, but I felt sad as well. We usually set the table nicely and eat by candlelight in a nonverbal acknowlegment that for an hour we will give each other our undivided attention. Bolting midway through and jumping into my car to drive the half hour to a ranch house in Elk River was just another kind of erosion.

The meeting was at a house that had belonged to Kristi Wrigley's father before he died. The address was Wrigley Road, and from my research, I knew that the Wrigleys had been some of the original settlers in the area at the turn of the century. I pulled in the driveway and dashed up to the door a few minutes late. I knocked softly, was admitted, and I was about to tiptoe into the living room where I could hear the meeting in progress, when a woman stopped me beside a little table and asked in a whisper if I would please sign a guest register. Glancing at my signature, she said, "I'm sorry . . . I should know you . . ." fetching for my name so she could place me in one house or another within the valley.

"Well, actually, I don't live here. I'm writing a book on Headwaters, and I simply came to . . ."

"I'm sorry," she said abruptly. "You'll have to leave. This meeting is only for residents."

"But I was invited by Ralph and Nona Kraus," I explained as she ushered me toward the door. We stood outside in the night air of the rural valley as she completed her dismissal of me.

"We are experiencing a lot of suffering and confusion, and we need to work this out among ourselves."

I tried to argue further, but I was later to learn that this was Kristi

Wrigley, a tough rancher who had wrestled with far greater intruders than myself.

By the time I got home, dinner was over. Kristi called the next day and apologized. She had talked to the Krauses and felt terrible. But I understood. I know what it's like to have the soil washed away from one's roots.

I saw Kristi again a few days later at one of Doug's slide shows. She was talking to a woman in the seat next to her, telling her that she wasn't really an "environmentalist" and she really didn't normally go in for this sort of thing. After the slide show, she was a different person. She had not seen the clearcuts from the air. Until she saw Doug's photographs, she had not known the magnitude of the problem that was making Elk River so silt-laden.

After the slide show, I reintroduced myself to Kristi, and she began apologizing again. I told her that I understood and asked if I could come see her sometime at her apple orchard.

"Yes, please." Suddenly she dropped her veneer of hard-bitten rancher and revealed her despair. She was simply a woman raising two children on a family ranch that was under siege. But there was no "family," at least no "pa" and ten gigantic brothers defending the homeplace. There was simply Kristi and her teenaged children, a lot of pride, and memories of an orchard that had served the public for generations. When I called Kristi's house to arrange a time to see her, her phone message informed me which apples were currently being picked and when they would be available for purchase at the ranch. I recalled my excitement as a child at going to just such places, where traditions persevered. We need more Kristis. We don't need to make life difficult for those very people who keep life rich.

CHAPTER 34 KRISTI

DAVID and I have just returned from visiting Kristi Wrigley. John Humboldt Gates's book on the Elk River lumber town of Falk, *Falk's Claim,* lists a whole column of Wrigleys. As we arrived on the narrow one-lane gravel road, Wrigley Road, that ends at Kristi's apple-packing shed, a young man was just getting out of a van with Oregon plates. Kristi walked out to greet us, and the man, who appeared to be with us, asked where the remains of Falk were located.

"It's gone," Kristi replied. "It was dismantled, torched, bombed by the lumber companies to keep the bottle hunters out. It's gone."

Having studied the map in the front of the book, I knew that, standing in Kristi's barnyard, we weren't far from the location of the vanished town that materialized at the turn of the century, as the Elk River Valley was having its forest stripped for the first time. Now the forest is being stripped again, only this time that harvest includes the old-growth that was once too distant to log economically. On my way up the peaceful valley with its hayfields and nearly hundred-year-old second-growth forest, I dodged logging trucks. Above the ranch, I saw that clearcuts were creeping over the hill with their mangy-dog appearance, tearing open the comforting illusion of wholeness which, each fall for decades, visitors have enjoyed when they come to the ranch to buy apples.

Kristi politely moved the man from Oregon on his way, accustomed to protecting her work time, and we made our way through a construction zone into Kristi's house, which was having its roof replaced. As we entered a large utilitarian kitchen, it was apparent that in preparation for the roofing job, the attic had been emptied, its choicest contents finding their way downstairs: a framed family tree, an old doctor's bag, a small iron lamp shaped like a little coach to illuminate a child's bedroom. After a lunch of home-canned albacore sandwiches, which Kristi's teenaged son and daughter finished preparing and served with remarkable helpfulness, I went upstairs to use the bathroom. As I passed the bedrooms, I noticed that handmade quilts covered each bed with the unselfconsciousness that characterizes houses that have sheltered generations of quilt makers. "Flying Geese" or "Grandmother's Flower Garden" were simply the coverings at hand in a place where history still wandered at will, as it had through Kristi's lunchtime conversation.

Beyond the bedrooms, daylight pierced the interior of the house beneath the sound of the roof being ripped off. This, too, was like Kristi's life. Her jaw is set, her hands are tough and cut, and her children shore her up with frequent hugs as she shoulders the family ranch without the man who left it to her. Mention of "Daddy" was sprinkled throughout her stories.

"Do you have a goal for the number of apples you want to pick today?" he still asks Kristi quietly from the grave.

"He never pressured. He simply asked. I still remember the day I picked fifty boxes," Kristi reminisced as we headed outside, down the road, and then entered tall grass on our way to the river. "I've got to *mow* this," she interrupted herself, brought into the present by every rancher's ready disgrace—grass in need of cutting. "I've got to mow all around those trees." We looked down on new acres of orchard, the rounded trees alive with promise.

"They haven't borne fruit yet," Kristi said flatly.

She is very concerned that the Timber Harvest Plans filed by MAXXAM for eight hundred acres of clearcuts upstream from her ranch include notice of intention to spray herbicides. The old Pacific Lumber did not poison its land. This is a new onslaught on the Wrigley ranch.

"But we take our *drinking* water from the river," Kristi told the biologist from MAXXAM. "He was very nice," she went on. "I *liked* him very much . . ."

"Kristi, if you owned a lumber company and wanted to spray herbicides, wouldn't you put your most likable guy in charge of that PR?" I asked her. Kristi is no fool, so obviously this man had been *very* nice. It is remarkable how most of the biologists working for MAXXAM and the agencies all still know what is right, while they are being pressured from above to abandon their own training and morals.

"He suggested we dig a well. I said to him, 'Dig a *well*? We've taken our water from this river for almost a hundred years!' The well water in this valley smells of sulfur and turns everything brown."

I looked at her beautiful daughter and across to the young orchard, and thought of my conversation the day before with a woman named Patty Clary, of Californians Against Toxics, and emergency room doctor and activist Ken Miller. Over lunch we had studied a list of the herbicides that MAXXAM is using and a map showing what areas had already been sprayed. Later I sat down alone with some information that Ken had lent me, much of it printed on the backside of paper recycled from his practice. On one side were notes on healing and on the other were prescriptions for death.

MAXXAM / Pacific Lumber, the largest user of herbicides in the county, has been using herbicides for only the last few years, to control the broadleaf shrubs and trees that attempt to revegetate their clearcuts. They use Atrazine, Oust, Garlon, and Roundup. For fifty years before the takeover in 1985, PL rarely clear-cut. MAXXAM, in order to pay interest on its debt, has increased clear-cutting until now virtually all THPs are for either clearcuts or two-stage clearcuts, that is, thinning followed in just a few years by clear-cutting. One of the shocking facts of herbicide use is that MAXXAM has no requirement to notify nearby residents of dates, times, or location of spraying. There is no requirement for direct monitoring of environmental or water quality. A company such as MAXXAM is only required to self-monitor application, and report, after the fact, that herbicides have been used.

Herbicides do not stay put where they are sprayed. They drift on the wind; stick to soil particles and move into the water supply with erosion; and are transferred by the urine, feces, and carcasses of animals. Atrazine and Roundup are estrogen mimics. Estrogen mimics are thought by many scientists to play havoc with the reproductive systems of animals, causing cancer, impotence, and sterility. Sublethal doses of Roundup can kill riparian or streamside vegetation and impair swimming in fish. Oust can retard fruiting in trees. As roots of shrubs sprayed with herbicide decompose, the resultant erosion plagues an area for years after the plants were actually killed.

We ducked into the riparian zone along the river, tall alders dappling the ceiling above us with that delicious color known as "sap green" when it comes from a paint tube.

"Look how high the river came last winter," Kristi's daughter said to me. We stood on a bank fifteen feet above the river, yet she pointed to a branch over my head whose covering of lichen and moss had trapped mud on its way downstream in last winter's floods.

"When I was a child, the bed of this river was all gravel," Kristi said, looking down. "We could count on seeing salmon spawning."

"Really?" her daughter asked. "I've *never* seen a salmon spawning. I've only seen one salmon, and we were mad because an otter had killed it."

"They used to fork out the salmon with pitchforks, there were so many here," Kristi went on. "Look at this mud! It's clay. You can't walk down there anymore, it's so slick. It's flooded into the orchard, and when it gets wet and we're out pruning, we slide and it sticks to everything."

I knew from recent experience what that was like. I had gone to Strong's Creek on the edge of Fortuna with students from Fortuna High who are monitoring this creek, which flows out of MAXXAM clearcuts. The silt formed a deep, trapping paste that might suck our rubber knee boots down like quicksand or spontaneously release them at any moment, landing us in the mud.

"Finding mud used to be a treat," Kristi continued. "When we were children, we explored the river and every now and then we would find a place where a bank had given way. But it would occupy just a small area briefly, and then it would wash away. This mud is everywhere, always."

I recalled how at Strong's Creek I had spotted a small tributary which came from an area that had not been recently logged. It curved into sight, its banks ferned, its bed cobbled, until it joined the main water course, where all its delicate variety was suddenly buried by silt. The high school students I accompanied, Kristi's daughter, any child who explores one of the many creeks that drain MAXXAM land, will have to make do with the monotony of silt.

I had come to see Kristi because the issue of private property is central to the fate of Headwaters, which, as a raven might fly, was about eight miles upslope. People contend, "It's Hurwitz's private property, and he can do what he wants with it." I agree, in theory. But in watershed after watershed below MAXXAM land, scenarios unfold almost identical to Kristi's. The fact is, MAXXAM is a bad neighbor to plant and animal alike, including humans. I came to see the main ranch house where Kristi had barred me from the meeting a few months before. After last year's flooding, the county assessor showed up and reassessed it at half its previous value.

As Kristi, her daughter, and I drove down the valley from the orchard to the ranch, I noticed that the public county road we traveled was rutted with potholes on the side used by those leaving the valley.

"That's from the weight of the loaded logging trucks," Kristi informed me.

"Why's the road wet?" I asked.

"To keep the dust down. Ever since they started logging in wintertime, the mud on the roads is terrible. When it dries, it blows everywhere, so they bring in water trucks to keep it down."

"When did the old PL used to stop logging for the year?"

"October, and they didn't start again till spring. Now they run thirty trucks three times a day all winter long."

"Thirty trucks?"

"Yes, in and out."

That computed to 180 trips per day on the narrow road. Even if Kristi had made her count on an unusually active day, this still suggested an overwhelming change in the local quality of life. Later, a little disbelieving, I checked this figure with a friend of mine who lives in the Carlotta Valley, and he said that he once counted fifty-five logging trucks an hour using two-lane Highway 36. Indeed, as we drove down Kristi's road, two loaded trucks followed us out of the valley.

"When people come for apples, they have to dodge the trucks constantly."

Again, we were talking in that gray area of "degree" and "quantity." It's not that the old PL didn't log, but it didn't log at this *rate*; it rarely clear-cut; it did not haul logs during the winter rains; it did not use herbicides; it did not "salvage" downed logs and snags from the forest, nor pull them from the river bars to be milled in a desperate scrounging for wood like a glutton licking its plate.

Pacific Lumber/MAXXAM owns approximately 210,000 acres. In 1981 they filed 13 THPs for the entire year, encompassing 1,624 acres. In 1988, three years after the takeover, they filed 53 THPs encompassing 10,436 acres. In recent years, approximately 100 THPs encompassing approximately 10,000 acres has been the norm, including hundreds of acres of virgin old-growth forest.

At the ranch we walked back down to the Elk River.

"The channel is filling up, isn't it?" I asked flatly.

"Yes. The water comes up more quickly but it takes longer to leave." It is fairly common knowledge in Humboldt County that a major flood has occurred every ten years, mid-decade, since 1955—1955, 1964, 1975, 1984, 1995.

In the winter of 1996, however, the house was flooded three times. El Niño? Yes, to some degree. But all the more reason for an upstream neighbor not to be clear-cutting steep slopes, clogging a river with gravel and silt.

"The house tolerated taking in water every ten years," Kristi told me. "It used to drain from the house in a few hours after a rain. Now it takes days. Last year my neighbors came without my even asking and helped me dig out. But what am I supposed to do with this house? In April the assessor showed up and cut the value of the house in half. When I went to get insurance, I was told I could no longer insure the house. I can now only rent it to someone who is willing to move in with the strong likelihood that they will have to vacate on short notice at least once each winter."

We walked inside and, as often happens, I was stunned by the private world that can exist within the confines of fairly modest exterior walls. A long room with a high-beamed ceiling and old-growth Douglas fir floors occupied much of the length of the house. It radiated family pride and integrity. Its design appeared to be based on the assumption that this family would gather here often over many generations. Its hearth was clearly intended to be the soul of the ranch.

All looked well until Kristi began to point out the floorboards that had buckled, the antique table legs with roughened finish, the mud-stained blinds, the lifted linoleum, the hearth that now grows fungus. I looked at the shiny fir floors most likely laid by Kristi's ancestors and tried to imagine them deep below sticky gray silt after floodwaters subsided. I looked out the window toward the river, at the beautiful alders that followed its course through the valley, and I saw a way of life needlessly destroyed.

We walked out in front and stood watching a neighbor hay her field, sweeping round and round on a rotary mower. The field is dry now. Even Kristi and her daughter looked in disbelief at how nature has covered its scouring with vegetation.

"Look at those fences," Kristi said, pointing to the far side of the field, haunted by work. "Look at how the water has pushed the stakes and raised the soil level and how debris is still caught on the barbs. All that has to be repaired. Where am I going to find the time just to fix my fences?"

And more Timber Harvest plans are being filed all the time. Each THP has a name, most of them, like the names of nuclear power plants, insidiously pleasant. A new one that has appeared on the books upslope from Kristi, is called The Archery Range. Poetic, evocative, homey, it refers to a Boy Scout camp that the old PL had long rented out to the various troops of Humboldt County. I have heard countless references to the camp made by grown men who went there as boys. In one man's recent memory, 1979 to be specific, the Elk River was cobbled and pristine and vast forest surrounded the camp. Four years ago he went back to this camp with his sons and described it to me as still functioning but surrounded by clearcuts and pampas grass.

As I listened to his description and studied the look of loss and confusion in his eyes, I wondered whether he would ever encourage his sons to return to the camp as scouts. There is nothing that says that hiking through alien species over an eroded landscape can't be enjoyable, but deep down most people will admit that hiking through an ancient forest imparts a different awe. I wonder how many powerful experiences such as this we can take from children during their formative years and still raise individuals for whom reverence comes easily. What are the boundaries of private property? How can we be so brainwashed as to vehemently defend the rights of a corporation, while all around us the quality of life of real human beings is being callously diminished by that same corporation?

Western bleeding heart

CHAPTER 35 AN ACCIDENT

I STUCK my fingers into the moving blades of a lawn mower. I didn't believe that people did this anymore, now that power lawn mowers have been a normal part of daily life for decades. At the time of the accident, however, most of my body was covered with poison oak from our hike through Yager Camp, and I was nearly delirious from watching the massive patches of weeping, itching blisters spread out of control.

I am selling my house in Ferndale, and when the poison oak was at its height, my realtor called to insist that I must hurry down and mow the lawn before a particularly promising buyer came to look at the property. Why didn't I hire someone? I think for the same reason I didn't turn off the mower before I tried to raise the housing so the blade would ride higher over the tall grass. I was nearly out of my mind. Even when I heard the blade hitting my fingers, I didn't realize what I had done. It was not until I pulled my left hand away and looked that I realized that I had joined the ranks of those people who do stupid things like cut the ends of their fingers off. And it was not until I was lying in the emergency room that I realized with a jolt that I would have to type with one hand for at least a month if not months.

CHAPTER 36 HOUSTON

I HAVE just returned from meeting Charles Hurwitz, face to face.

When Ken Miller told me that he and Ellen Fred were going to go to the annual MAXXAM shareholders' meeting in Houston, I said, "I want to go too."

"You do?"

"Yes."

There was no question. Lewis Carroll hears that Alice, a figment of his imagination, will be appearing in Houston. Surely he goes. He goes to see that she is real, walks, talks.

The shareholders' meeting was held in the Houstonian, a hotel on the edge of the city, down a long drive, past a guard house, beneath the spreading oaks that shade the more affluent areas of Houston. Trees. I understand that Charles Hurwitz lives there at the Houstonian, beneath those trees. For so many months I have wondered to myself, what were the trees of Charles

Hurwitz's childhood? Where does he live now? Does he live where there are no trees, or where trees are simply token decorations, so that one does not have to be distracted by their presence or their absence? While marbled murrelets in Humboldt County circle, looking for trees that have vanished, I have wondered if the trees that vegetate Charles Hurwitz's surroundings are safe and secure. They are. The oaks of Houston are unforgettable. I cannot describe the exterior of Houston's plush art museum, or of its downtown skyscrapers, or even of the Houstonian itself. But I remember Houston's trees.

As several of us approached the hotel by car down a long, shaded drive, we crossed an AstroTurf track where I have been told Hurwitz runs to stay in shape. It wound beneath the oaks. Ellen told me the night before that in the little purple bag that hangs close to her heart from a cord around her neck she carried a piece of charcoal from a charred stump of an ancient redwood.

"I went to the track where Hurwitz runs and drew a line across the track with the charcoal."

"You *did?*"

As we drove by, I imagined her in the darkness drawing this line with the carbon of tree rings formed thousands of years ago. It would never occur to me to do this sort of thing. Sometimes I am amazed by the sense of ritual that is alive in people like Ellen. Where do they get their Druid blood?

A woman in matched sweats walked a small white dog on the track. They, too, would unknowingly cross the charcoal line.

Spacious grounds opened up, lavish flower beds surrounding a country-manor-like hotel. A line of valets leapt to take cars and park them.

"Let's park the car ourselves, so we know where it is," Ken said cautiously. We were walking into the lair of the opposition. Just as well to keep track of the car.

A large centerpiece with eucalyptus provided cover for a stuffed pheasant who peered at us through glass eyes as we entered. This was framed by a wide stone fireplace in the lavish sitting room beyond. We followed the flow of people in business suits toward one of the conference rooms and encountered a table set up for the scrutiny of our proxies. I used Doug's. He owns five shares so that he can attend these annual meetings, though he never has. Our proxies approved, we went through a metal detector, sensitive enough to pick up the hooks on my purse straps. Suspicion runs both ways.

Ken and I walked together into the conference room. Heavy plastic chairs were arranged so that the audience would face an impressive, multi-miked wooden podium, behind which was an entire glass wall which looked out into lush deciduous forest. I thought to myself, "Warblers." And then I began to survey the rest of the room, my eyes stalling on a table loaded with silver coffeepots and teapots, and beautifully arranged silver platters of pastries and sliced fruits. Suddenly Ken whispered, "There he is."

I looked up as Ken stepped forward to shake hands with Ezra Levin, a director on MAXXAM's board, who greeted Ken by name. Ken not only attended the shareholders' meeting last year, he met Levin a few months later for lunch on the West Coast. The conversation was cordial, familiar, and my head started spinning. If we all got along this well, why had I been writing this book in solitary confinement for the past five months? Had we only needed to meet and share beautiful fruit and coffee and pastries?

Meanwhile, my focus instinctively lengthened past Levin, sensing that the urgency in Ken's voice when he said "There he is" did not refer to Levin. There he was. Charles Hurwitz, pausing to greet shareholders as he followed Levin through the room. It was almost too much to expect. I had, in fact, anticipated only that I would study a distant, small figure in a dark suit on stage in a large auditorium. Lewis Carroll does not expect or necessarily even want to talk to Alice, only passively sneak a glimpse of her and verify that through some unexpected transformation she has been made flesh and blood.

Yet, just beyond Levin, rapidly approaching us, was Charles Hurwitz himself, straight off the cover of *Leaders*, a little more drawn, a little paler, a few years older. My modest expectations felt rushed by reality. Oh . . . you're getting older . . . as you cut these forests . . . it doesn't fully agree with you . . . it's taken its toll . . . the protests and all . . .

"Hello, Ken."

"Hello, Charles." They shook hands and bantered momentarily in that way that men do who have contempt for one another, and then Ken said, "I'd like to introduce my friend Joan Dunning."

I looked into Charles Hurwitz's eyes as I extended my hand. He is not a myth. He is a man. Quite simply, it takes only a man to direct the mass destruction of thousand-year-old trees. Only a man, cordial, practiced at killing time with people he does not care to know. He does not emanate warmth. He is not a humanist. I was shaking hands with a major force in the economic globalization of the planet. He is a well-oiled taker.

We shook hands and he said, "But you're so pretty . . . What are you doing with someone like Ken?"

This was the equivalant of banter between males. I was not moved to ask him, "Have you hiked alone in Headwaters? Spent the night there? Have you seen a spotted owl? Have you sat and called an owl in through the forest to perch above you and stare down at you? Have you ever drawn a fern?" I did wonder if after the months that I had spent writing about and visiting this man's forest, there might be some ghost of a forest in my eyes. But I really had no expectation that he would recognize it as his own.

Instead, Hurwitz told us a couple of stories that I imagine he saves for Humboldt County residents. He said that he went into a ma-and-pa store in a small town south of Scotia, and he asked the clerk, "What do you think of Pacific Lumber?"

"Oh, they *used* to be a good company," the clerk said.

"*Used* to be?" Hurwitz prodded.

"Yeah, until MAXXAM bought 'em out and brought in those spotted owls."

The other story was in a similar vein. I forget it though.

As I think back, meeting Hurwitz was like a point of stasis where virtually nothing happened. This is the intention. Each year he simply puts in an appearance before his shareholders and surveys his opponents. His shareholders watched silently, as I did. His more committed opponents poured their hearts out when they were given permission to speak. I never doubted that traveling to Houston was worthwhile. It was worth it just to lay eyes on

Charles Hurwitz. And it was worth it to see his opposition in peak form. But I have no illusion that when the board of directors went into closed session after the public meeting, that they did anything more than spend fifteen minutes writing down names and smirking condescendingly before getting on with business.

When we were asked to take a seat for the meeting, I sat in the second row with Ken and Ellen. The front row remained empty until Darryl Cherney strode straight down the aisle in a blue cowboy hat and sat down alone in the front row. Darryl is not particularly tall. His stature is conveyed through his words, not his physical presence. No sooner had he sat down than I became aware that his plastic chair was perhaps the only one in the room that squeaked. Later on, when I commented on the squeaking and wondered if Darryl were deliberately rocking it, someone told me that Darryl simply never sits still.

I watched for warblers in the treetops outside rather than make eye contact with the somber men seated within the massive podium, while Darryl kept an eye on everyone in the room.

Immediately it was announced that five people in the audience would be allowed to speak during each of two designated comment periods. The total of ten people would be drawn from a pool of the names of those people who wished to speak.

After the board of directors successfully suppressed opposition to the system outlined, the first five names which had been drawn from the pool were read aloud. The crowd was shocked when the list read like a who's who of Hurwitz detractors. I would give anything if I had been allowed to tape-record the whole meeting, because the minute the names were read, I felt as if I had slipped into the audience of a superbly written and well-cast drama.

Most memorable was Darryl Cherney, in his blue cowboy hat, who walked up to the microphone, faced Charles Hurwitz squarely, and said, "My name is Darryl Cherney, and I am a MAXXAM shareholder. Under Security Exchange Commission regulations it is my legal duty to inform you that I have quietly purchased five shares of MAXXAM stock, and it is my intention to take over your company.

"But, unlike your takeover of Pacific Lumber, I don't intend to rip off a savings and loan, loot the MAXXAM pension fund, or bribe your board of directors. However, I do not plan to tell you my methodologies any more than you told Pacific Lumber Company how you were going to take them over.

"If you would like to have at least a clue, however, you should be made aware that we have issued a twenty-five-thousand dollar reward for your arrest, conviction, and incarceration. If you'd like to learn more, you can look up our Web site at www.jailhurwitz.com. Thank you."

There were lots of people who communicated on MAXXAM's terms. But what they said made no difference, perhaps they simply bored MAXXAM's directors. Their suggestions and objections were not discussed and considered in "meaningful dialogue." No one on the podium suddenly sat up straight and said, "Actually, you've got a point there. Let's discuss it and see if we can agree on some changes in our policies."

As I listened to Darryl, I imagined him regaining consciousness after the bomb blast, both eardrums blown out, his face bleeding, his left eye temporarily blinded, trapped in the car next to Judi Bari, who was near death and screaming. Darryl's speech was a sobering display of personal responsibility, made all the more profound because personal responsibility is precisely what is missing in the MAXXAM Corporation.

Darryl had a chance to speak again. This time, even though we had scarely ever spoken, he echoed sentiments that I have felt. I cannot exactly remember his whole speech. I phoned him, and he can't remember it either. But the gist of it was, "You think you own the Pacific Lumber Company and all of its redwood forestland, but the truth is that you have never walked this land. You don't even know what you think you own. I would suggest going out there for at least three days to get to know your own forest."

He then launched into an articulate appeal, most of which I can't remember, except these words, "Mr. Hurwitz, you were a boy not long ago..."

Ken stood up and suggested that MAXXAM might be more effective if its board contained some members who actually lived in Humboldt County and were enduring the consequences of the widespread logging.

I chose to remain silent. I was there to report. I wanted only to listen and feel. To muster words would have been to throw my mind into conversation.

Ellen went to the mike and instead of speaking sang a song. I cringed at first, looking nervously back and forth between the men and Ellen, who sang a capella, her hands at her sides. To my surprise, the men actually seemed to listen with appreciation. At the same time, I wondered why any of us was attempting to entertain them. The poignant song asked, from a redwood's point of view, "Who will cast my shadow when I am gone?"

After the meeting we milled around, eating the pastries and fruit, drinking coffee. I edged in to listen to a conversation in which Hurwitz was saying, "It's my private property. It's my right to do with it as I choose." I abruptly broke my agreement with myself to remain silent and lamely began to describe Kristi Wrigley to Hurwitz. He listened to me patiently, with no expression. He knew there were Kristi Wrigleys. I was telling him nothing new. He gave me an appropriate hearing and then brusquely replied, "I don't know about this. You'll have to ask John Campbell." Of course, John Campbell, head of operations in Scotia, knows of Kristi Wrigley, but it makes no difference.

Hurwitz's son was there at the meeting, a young man in in his twenties, tall and capable looking. Ellen began talking with him and invited him out for a hike on his father's land. He said, quite sincerely, "Oh, my mother wouldn't let me."

"Wouldn't *let* you?"

"She'd be afraid of what you'd do to me."

We looked at one another, trying to imagine their fears. How little he knew about us. How much we could share about life, not its endangerment. We could all lie, predawn, in our sleeping bags and watch murrelets arrive to feed their chicks, each waiting on its own branch in the canopy above our heads for the arrival of dawn and the glint of that little silver breakfast, a fish fresh from the sea, passed to the next generation of a dying species. I could take him to my old-crone tree in Headwaters, and he could enter her sanctity and peer out on his father's forest and imagine its best use.

Redwood sorrel

CHAPTER 37 RETURN

I AM back—to the redwoods. I find a brief patch of sunlight and lean against the charred trunk of a giant. The sun will light here only momentarily, having found a hole in the canopy. I am by Calf Creek in Humboldt Redwoods, and the sound of cars comes only occasionally and distantly. Cars. Cars. Cars. I am back from Houston. I am exhausted by the sound and sight of cars. What if no one had saved these redwoods to take me back and hold me with their mass? What if I could not sit and watch this bumblebee work this oxalis flower? What if this creeper were not singing, "Trees, trees, beautiful trees"? What if the wind did not sing a song of time over my head? What if that tree did not squeak its trunk against that of another, and I could not lay myself prone on the forest floor and watch it sway in a wind I do not feel? I am here to tattle and have the trees soothe me. They do. Their long view makes my heart still.

I whisper "Thank you" to the Earth below me, into the graves of the men and women of the Save-the-Redwoods League, who are responsible for holding off the cutting of these trunks. With the crazed activity of Houston not yet flushed from my nerves, these trees that rise around me do not seem like a forest, but like standing Buddhas. They say, "Be." And they say, "Be still." And they draw to the surface grief I didn't even know I carried. Their fibrous bark, like a poultice, wicks away despair that I carry from the shareholders' meeting. Did we touch any heart? Was there any heart exposed? Did we sow any seed of doubt or hope? Did we have any meeting of minds, or did we crane forward only from one side?

For now, I am back, and these trees are still here and will be when my children feel despair and need stillness.

CHAPTER 38 PUBLIC RELATIONS

THROUGHOUT the past year, I have been asked, "Well, have you gone to PL and heard *their* side of the story? There's always two sides to every controversy."

This question instantly poses a problem, because it's hard to believe that Pacific Lumber Company still exists.

My answer is that I have refused to visit MAXXAM's Scotia headquarters only to be fed a prepared presentation of company policy by their PR person, Mary Bullwinkle, because I regularly hear other MAXXAM employees speak at community meetings that I attend.

For instance, at a recent meeting of the water board in Hydesville, homeowners in the Yager Creek drainage filled the meeting room to standing-room-only capacity. For months, like the people in Elk River, this growing citizen group has been studying the problems that are currently diminishing their quality of life due to clear-cutting and herbicide spraying, in order to ask the water board to take a stand on behalf of the community that elected them. Because the Hydesville town well lies directly downstream from herbicide spraying that has turned the bulldozed slopes an eerie red-brown, the residents, who have already experienced the degradation of their "viewsheds"—increased siltation, exposure to airborne herbicides, and mounting fear of landslides during coming winters—had logically laid out their request that the water board write a formal letter demanding that no herbicides be sprayed upslope of the town well. This seemed especially reasonable since, up until two years ago, PL never used herbicides in the valley during nearly a hundred years of sustainable logging.

Also up until the past year, few people had ever attended the water board meetings, and opinions of both the members of the water board and residents have been based to a large extent on information presented by the one local daily newspaper, the *Times-Standard*, and two local news stations, both of which are heavily influenced by the timber industry. Relatively new residents who have moved to Hydesville from more affluent, better-educated parts of the country have received a rather sobering education in small-town politics over the past year, especially since many of them have their life savings tied up in their land and houses, but they still believed that the water board and MAXXAM would agree to their well-documented request.

When the townspeople presented their request, however, the water board put up hours of resistance until one frustrated resident finally turned directly toward MAXXAM's resource manager, Tom Herman, who sat silently in the audience, and demanded, "What about our water?"

Tom shrugged, smiled a lazy smile, and quipped, "Don't flush your toilets."

The comment was so incongruous and in such bad taste that no one in the room knew how to respond. The residents looked at one another, stunned, disgusted, and the comment stood starkly as a typical example of how residents' concerns in all the watersheds are ignored or even ridiculed. After several more meetings, the water board finally agreed to write the requested letter, but, as of this writing, it remains "stalled in draft form." Meanwhile the trees fall and herbicides are sprayed.

After countless other meetings in other watersheds throughout this past year, I cannot imagine going down to Scotia and deliberately exposing myself to the kind of PR I see in MAXXAM's television and newspaper ads. Instead I spend my time interviewing townspeople, such as an old-time resident of Hydesville whose ranch exists intact amid the encroaching deluxe ranchettes. As we walked across his fields and stood beside a marshy area from which he has excluded cattle in order to allow the area to return to wetland, he looked up at the surrounding slopes with a sad, perplexed, and bitter refrain that I have heard repeated again and again.

"We had no problem with the old PL. They were our neighbors. Many of my friends worked for them. PL was a member of our community."

CHAPTER 39 JEAN AT THE PIANO

I HAVE been out of town teaching a drawing seminar in Oregon, and now I am home again working for two weeks at a nearby arts academy for children. I feel remote from the forest. Doug called to say that he was hiking out to Shaw Creek Grove with some reporters, and he wondered if I could go along. But I cannot go. Last night I missed the Taxpayers for Headwaters meeting because I was at parents' night at the art academy, seated in a circle with parents, who were requested by the director, Jean, to describe what they most love about their children.

I am teaching those children to draw each day, preparing them for adulthood, arming them with the arts. The children, from different economic and racial backgrounds, are extraordinarily talented. At any moment one can hear several pianos from different ends of the house, the sound of a violin coming faintly from a distance, or from the porch, the sound of xylophones. Children beg to go to this place each year because they are treated with dignity.

However, while I have been focused on the children, the trees in Headwaters have been spray-painted with large white numbers and dated. An orange string now runs through the forest. One doesn't have to cut down a forest to defile it. I imagine going back to Headwaters and seeing the trees spray-painted. I imagine being hung up in orange string that runs insidiously through the ferns. I think of the delicate moss that one does not want to set one's boot upon. I remember the holy feeling of the forest. Would we callously walk into Chartres, or St. John's Cathedral in New York City, or Temple Emanu-El in San Francisco, and spray-paint walls or statuary, and think for a minute that we do not diminish the experience of the person who goes there, broken, in need of repair, desperate for a moment of restoration?

I understand, after questioning a few people, that the paint is probably put there by surveyors for Clinton's "Deal." Yet the Deal protects too little, just the Headwaters Grove and the Elk Head Springs Grove, but not the surrounding watersheds. It protects "trees" as a shrine, but not a whole ecosystem. Now they take those very trees and spray them with numbers, as if we did not value their bark but only their board feet. True, the voles will probably not mind if white numbers grace their trees. A murrelet would still nest in a tree with a large white number on its flank. A spotted owl would perch there. When a tree dies and falls, after having stood as a snag for hundreds of years, and then slowly rots over hundreds more, the mice that use its curving underside as cover, running from log to log in a familiar, owl-safe highway through the forest, will not care if white spray paint ever defiled its bark. The carpenter ants and boring beetles who turn the trunks back to soil will probably not mind if trunks were ever sprayed. But I do.

Is the struggle over Headwaters a religious war? Perhaps. The religion of materialism and the religion of the intangible are duking it out. One is a heavy-footed warrior and the other is a ghost. Yet the ghost, of soul, of spirit, of the meaning of life itself, will not go away. It asks, what is the reason that we live? Is it only to be "housed"? Is it only to have a redwood deck? Is it only to have redwood shingles or siding? How many parents have experienced the anguish of providing a happy home, only to see a beloved child slip across a rot-proof deck and away in search of escape from meaninglessness?

Yesterday at the academy, the beloved director, Jean, who has been my daughter Suzanne's piano teacher for the past seven years, walked up to the piano after lunch and sat down to play. She rarely does this in public. There was no audience. Children were in the kitchen scraping their plates and putting their cloth napkins in the hamper. As Jean played *Moonlight Sonata*, however, weighed by a personal concern that I knew was haunting her, the children, one by one, stole in and seated themselves around her on the rug. She did not seem to notice they were there, but kept playing, rocking her body slightly. When she was done, she laid her hands on her lap and bowed her head. There was silence, and then the children began to quietly applaud. Slowly Jean lifted her head and turned to look at them. She didn't say a word but gratefully met their eyes with an expression that conveyed the power of music to console. I thought to myself, if only there were more of this . . . more soul, less materialism. If only we slowed and invited children into our adult hearts, we would weigh less heavily on the Earth.

CHAPTER 40 THE ACCIDENTAL ACTIVIST

I WAS flown to San Francisco for the day last Friday to read at Temple Emanu-El, the largest Jewish synagogue in Northern California. The synagogue has organized a "Call to Action," to inspire a multidenominational uprising against Hurwitz's destruction of the old-growth redwoods. When I was introduced by Rabbi Pearce to the audience, he commented that the title of my book could be *The Accidental Activist*. Most of us involved with Headwaters feel as if we stumbled into this. I watch new people, a little dazed, act as if they have fallen down a rabbit hole and will soon escape. Yet, as the months have passed, I have felt a rising and deepening sense of destiny. I now refer to "the forest." From its hillsides up Freshwater Creek, and Elk River, and Salmon Creek, and Strongs Creek, and Yager Creek, and Cummings Creek, it directs us.

The other night my kids and I went with Chuck, my partner/advisor/at-home-editor/general handler, to a restaurant for dinner. When we got into the car, we tuned in to the middle of a radio interview. The person being interviewed described how more and more people are being called into the service of the environment to do very precise tasks for which they seem to be perfectly prepared by past experience, even by what they considered to be past "mistakes." When we stopped in the restaurant's parking lot, I couldn't bear to leave the car. I had to keep listening. Chuck brought me out a glass of wine, and then he and my children went inside to order. They are used to doing without me.

I did not even know whom I was listening to while tears of recognition acknowledged the tension and responsibility I feel in my work as a writer. After a station break, I learned that the person being interviewed was the author of *The Celestine Prophesy*.

I was surprised. Though I have found myself questioning its main theme repeatedly, that is, whether we attract through our own energy the events of our lives, I didn't like the book as a whole. Yet live, on the radio, I found James Redfield hypnotic. He was telling my story. He echoed my sense that

Red tree vole

Americans are, one by one, being called forth to act on behalf of the natural world. In a way, this is no surprise. As things get bad enough, people are bound to respond to some degree. But the actualization of this impulse feels collectively spiritual, rather than simply tardy.

At my reading at Temple Emanu-El, I stood before approximately one hundred of the foremost rabbis, priests, ministers, and other religious leaders of Northern California, people who make their livings doing in person what I generally do on paper. I began by saying that when I was eight, I told my father that I wanted to be an "ecologist." I wasn't exactly sure what this meant, but I had seen the cover of the Time/Life book called *Ecology*, and something about the entwined crabs spoke to me. When I was ten, I told my mother that I wanted to be a minister or a nun. Since I wasn't raised within any organized religion, my mother looked a little perplexed. I was too. What god would I serve? Becoming a nature writer appears to be the best way to marry the two impulses.

This said, I began reading. I am testing my wings as a reader. It is one thing to read aloud to friends and have them cry, and quite another to read to a room full of professional ministers and rabbis and have many of them confess later that they were moved to tears during the reading. Afterwards people touched me, hugged me. This was how I knew that I had been of use to the forest, that, like a fledgling bird, I am capable of taking off, flying, and perching in the hearts of others. This is all I ask.

CHAPTER 41 SLASH BURNING

BECAUSE of the rain, the last three mornings have been like autumn, stunningly clear after a summer of gray mornings. I got up and looked out my bedroom window at the sun coming over the ridge through the branches of a redwood in my yard. Then I came downstairs ready to work another day. I pulled open the blind on the south-facing window in my studio and stopped. "Oh . . . there is a cloud in the sky after all. Funny . . . a single cloud . . . it seemed so clear out . . . " I stared at this cloud for a minute, pondering what it meant about a day that I thought was cloud-proof. Then I suddenly noticed it had a very slight red tinge.

Of course . . . the first rains . . . they clear the air, but they also lift the ban on slash burning. I was looking at smoke rising from MAXXAM *land fifty miles to the south. Depending on the wind, by afternoon this smoke might cover the whole sky. Certainly for the people of Elk River, Strongs Creek, and Cummings Creek it would.*

Theoretically, slash burning reduces the amount of dead wood and small trees and shrubs following a clearcut, minimizing fire danger and permitting replanting. The beauty of a redwood forest, however, is that it resprouts without slash burning if it is properly thinned. Slash burning fertilizes the soil in the short run, but it robs the soil in the long run because nutrients are released rapidly into the soil and the air, instead of being incorporated by the soil community slowly over time. Because the old PL did not practice wide-scale liquidation logging, slash burning used to be minimal. Perhaps this was less profitable in the short run, but it was more sustainable.

I took my children to school and as I drove home, heading north, away from the spreading smoke, I again savored the clarity of the morning. I always forget how much I like autumn until the first rains come. During July and August, I feel increasing dread, and then with the first rains, I feel cleansed and open to whatever winter has to offer. Fall is a transcendent time. If the death of the year comes so beautifully, then it must be right. The mere passage of the seasons is one of those simple pleasures that is enough to sustain the soul.

But then I wonder why Charles Hurwitz, at this very moment, is sending up a monument to his own insatiable greed in the form of a fanning plume of reddish smoke that threatens to envelop the day of thousands and thousands of citizens. I look south and think of my friends in Fortuna, Ferndale, Elk River, and imagine them glancing out their office or kitchen windows and suddenly feeling unexpected gloom. What happened to this bright morning's passage into a crystalline afternoon? Why has the day turned common? They may not actually walk out to look up at the sky, but they will feel the loss. Instead of transcendent, they will feel mortal. We all will.

CHAPTER 42 PLUMMETING

IT is the day of the Headwaters Rally for 1997. When I went to the rally last year, it was simply to sell my marbled murrelet T-shirts. I did not even listen to the speakers and entertainers. I just stood by my card table alongside the road and sold shirts. I wanted people to see a murrelet, to read about one. I wanted them to own the murrelet as part of themselves. Occasionally I see people on the street wearing one of the shirts, and I feel I have enabled the little murrelets to be ambassadors for themselves.

Today is the traditional date of the Headwaters Rally because September 15 is the official end of the marbled murrelet breeding season for the year. In human terms, it is the day Bonnie Raitt, Bob Weir, Jerry Brown, and other celebrities perform for thousands, perhaps in the rain. But for the murrelet, it is the last day that MAXXAM must hold off operations in the old-growth stands that are home.

Chuck came to me a few weeks ago with a piece of paper that, as I write, is still confidential. It shows the plummeting of the marbled murrelet, but not out of a redwood. It is a graphic projection of the bird's population over the next thirty years. Chuck tells me that it is considered "the best possible science," the basis upon which law on an endangered species must be based. For months scientists have hovered over computers, feeding them information and then waiting for them to speak like oracles. I sense that the biologists already knew in their hearts what that last computer would finally say as it finished tallying and projecting. They have been on the ocean in boats. They have seen what any Indian in a dugout redwood canoe could have seen. The little black and white birds that used to carpet the surface of the sea, rising en masse as waves passed beneath them and dropping en masse into the waves' troughs, are gone. The trees that used to carpet the wavelike slopes inland are gone. The rivers bleed soil, staining the ocean in greater and greater fans of brown. The marbled murrelet may go extinct in California. There simply may not be enough birds left to sustain the population.

Chuck has spent the past ten years fighting for the murrelet, lying under trees watching it in the first light of dawn, teaching others to recognize it in flight. He did not know when he questioned his first THP in the Carlotta Valley shortly after the takeover that he would wind up caring so much about the murrelet. He did not know that he would stand with a paper in his hand holding back tears as he told me that soon there might not be any more murrelets in the continental United States.

What is out on the surface of the ocean right now—as I write, as Bonnie Raitt makes final plans for what she will wear or what she will sing or say; as Bob Weir hears his alarm and rolls over for just a few more moments' rest; as Darryl Cherney ends a fitful night listening to the rain and wondering if there are enough Porta-Pottis for unknown thousands of people who are scanning the clouds at dawn planning their day—what is out there is a dwindling population of adults, and virtually no juveniles or chicks. We are losing.

CHAPTER 43 SMUT

IT is hard to imagine that *anyone* will be at the rally. It has been raining all night. MAXXAM has been running misleading ads nationally that make the Deal appear to be a "done deal," though the issue of appropriations is still being discussed in Congress. As further discouragement, the *Times-Standard* ran a front-page headline, "Headwaters Rally Forming," above two color photographs. The one on the left shows a well-groomed, perky young woman in shorts who holds picket signs for the residents of Carlotta to stick in their lawns. The signs read, "THIS FAMILY SUPPORTED BY TIMBER DOLLARS." "WE SUPPORT THE TIMBER INDUSTRY." "PRIVATE PROPERTY, NO PARKING, VEHICLE WILL BE TOWED AT OWNER'S EXPENSE." The caption reads, "Kathy Mason, clerk at Hydesville Mitey Mart, shows some of the signs offered to residents preparing for the influx of thousands of people on Sunday."

The photo on the right shows a loose circle of men in a redwood forest, some sitting at picnic tables, others kneeling on the ground. Most are clean-cut, but one is bearded, one is barefoot, and one was caught with an awkward Mansonesque expression that perhaps made even him grimace at the sight of the photo. The caption reads, "Tim Ream, right, of Eugene, Ore., discusses media relations with the first of 12,000 protesters expected at Sunday's rally." The message of the photo is simply "scruffy," "scary," and "out of town." Tim Ream is the young man who held me motionless a year ago while he described the cutting of an old-growth redwood at the rally in Arcata. From this photograph in the paper one might think he is not a genuine, articulate, and responsible American citizen, but rather the ringleader of a shifty band of troublemakers.

As I write, I glance out the window, and suddenly thunderheads roll apart and blue sky, in this chronically overcast area, reaches into my soul and whispers thoughts of hope. I learned last night, just before I went to bed, that rally organizers have opted to move the location of the rally from Fisher Gate to the small community of Stafford, south of Scotia. Stafford is the scene of a gigantic mudslide beneath a PL/MAXXAM clearcut; most of the community was wiped out by it. Rally organizers feel Stafford is a fitting backdrop for the rally. With luck, the members of the media will shoot footage of houses that are still filled with hardened mud, and wonder aloud why nothing has been done about them. They may even walk up and look into one of the houses, as I did the other day, and see that the fiberboard ceiling tiles have been removed and stacked by a resident who had to work on his or her knees in the limited space between the year-old landslide and the ceiling. Perhaps the members of the press will be touched, as I was, by this frugality in the midst of utter ruin. I hope that having Stafford as a backdrop will give rise to discussion of the efforts by the local people in all the watersheds to protect their quality of life.

It is my fear, however, that all of this concientious work will be overshadowed by police mayhem such as occurred at the smaller, nightmarish November 15 rally last year. Two days ago, I saw squadrons of motorcycle policemen from the San Francisco Bay Area heading north through the redwoods on Route 101, riding in precise formation. In the *Times-Standard*, the Humboldt County Sheriff's Department has already announced that anyone who "pushed through police lines would risk a felony charge of assault on a peace officer." This seems inflammatory. I do not recall any demonstrators behaving other than peacefully at the rallies over the past year.

In an hour I will leave. I am taking David and Suzanne and one of her friends with me to the rally. On the radio, attendees have been urged by organizers to make this a family affair. Why not? We are simply going to protest the stripping of the land of the ancient redwood forest. Why shouldn't my children be included? Why should I feel fearful to take them along? I resent that police conduct on November 15 has conditioned me to expect unpredictable behavior by law enforcement officers.

An image suddenly comes to my mind from the exhibit of delicate glass models of plants at the Peabody Museum of Natural History near Harvard University. It is my memory of a replica, indistinguishable from the real thing, of an ear of corn infected with smut. I take out my encyclopedia, and I look at a photograph of an ear of corn infected with smut. The affected kernels have retained their integrity as kernels, that is, they still have their outer cell walls intact, but they have enlarged grotesquely and turned brown because they are bulging with fungal spores. When I speak about Headwaters to friends in other parts of the country, they are appalled by my reports about police behavior, the bias of the local press, and the extreme reactions of many of the local people. My friends, at least with regard to this issue, live in the adjacent, yellow, wholesome kernels. We in Humboldt County are living in a brown, furry, nightmarish kernel that is expanding.

Douglas squirrel

CHAPTER 44 HEADWATERS RALLY, 1997

MY desk is filled with newspapers. The rally is over. The photographs and the words piled around me bear only scant resemblance to our experience there.

We arrived at exactly noon, the hour at which the rally was to begin. Immediately past the PL mill town of Scotia, one comes to the first freeway sign for Stafford, "STAFFORD EXIT ONE MILE." We still had to cross a bridge over the Eel River, round a bend, and we would be there. I found myself thinking . . . "If I were not writing a book and if I did not know what I have seen over the past year, I would not have come . . . I hate this . . ." Most of the "straight" people I know, especially those with children, were scared off by the media. They used their children, their jobs, their responsibilities, as legitimate excuses.

"I would like to go but . . ."

". . . but," I finished their sentences for them as I approached the Stafford off-ramp, "I am, quite frankly, afraid."

We came to a flashing orange sign that read, "SLOW. EVENT AHEAD." Then we rounded the bend, and we all sat up straight and stared. Before us the entire freeway, southbound and northbound, was occupied by lines of squad cars, lines of policemen in riot gear, and orange cones. My eyes struggled to assess the danger and also not hit any orange cones. The last thing I wanted to do was have a traffic cone tangled up underneath my car with all of these policemen standing around just looking for a problem. I considered going past the exit, and saying to the kids, "Oh, let's just skip this and go down to Humboldt Redwoods and swim in the river." They would have been surprised, but they probably would have gladly accepted.

Instead, I turned my wheel slightly and committed myself to entering the off-ramp alongside a line of identical police cars parked bumper to bumper. "Here we go . . ." I breathed.

Uniformed security guards alternated with T-shirted "Peacekeepers," all of them aggressively directing me to hurry off the ramp and under the freeway past more throngs of policemen in riot gear. "Oh boy. Here we go," I said again as we were funneled into a single file to bound into the grass of an open field. The number of cars already parked was shocking, and behind us we could look up and see cars backed up along the freeway. I eyed the trees beyond the field, looking for escape routes we might take on foot if the demonstration turned violent.

"We can always head for the trees, cross the river, and come out on the Shively Road," I comforted the kids.

"Yep," David agreed, scanning up the clear-cut hillside.

I turned off the engine and everything was suddenly quiet. I was startled by the contrast of the serenity. I was comforted by overhearing the people in the car next to us leisurely gathering up their water bottles, lawn chairs, and lunches. They spoke softly:

"Do you think we'll need one water bottle or two?"

The breeze blew through the leaves of a nearby stand of alders. The clouds, after the rain, were magnificent, clean and white and towering. In the distance I could hear a voice coming over a microphone. We gathered up our water bottles and umbrellas, in case of a shower, and our picnic lunch, and then we joined the crowd crossing the field toward the road on which we had entered.

We turned left along a lane shaded by second-growth redwoods. Driveways that led to neat, modest houses turned off at regular intervals. The people in the crowd talked softly. I normally walk faster than most people, but I realized that no one was passing other people in the line of walkers. We all moved at the same peaceful speed, as if this were rehearsed, occupying exactly half of the road to the left, while cars passed us every now and then, going the other direction. No one yelled, no one even talked loudly. We passed two horses who hung their heads over the fence watching the nearly silent passage of thousands of people down their usually empty lane. Even the relaxation of the horses struck me as unusual, almost dreamlike. There were no police anywhere. They were all back at the freeway—scores of squad cars, countless motorcycles, three prison buses, and hundreds of officers.

After a half-mile walk, we reached the rally site. No uniformed policemen were here either. A crowd of people, later estimated at six thousand, were in the process of seating themselves on the ground, each person calculating his or her own approach to sitting among thistles and mowed blackberry canes. For the next several hours we listened in peace to the speakers and the singing. Longtime residents of Stafford welcomed us. Kristi Wrigley thanked us for showing our support. Doug stood up and described the changes he had observed during six years of photographing the forest.

Someone took the mike and said that the police had closed the southbound off-ramp to any more cars. I felt a quick chill. Then Bonnie Raitt sang a beautiful song about the redwoods. Former California governor Jerry Brown spoke eloquently against MAXXAM. Cecilia Lanman, one of the cofounders of EPIC, who has been that organization's primary force, its rudder and its keel for twenty years, described EPIC's role in the preservation of Headwaters. We got word again that the police were not allowing cars into the rally site. Grateful Dead drummer Mickey Hart led participants in a rhythmic heartbeat of drumming. Student rabbi Naomi Steinberg sang a prayer with her young son. Chuck joined us as we ate our picnic lunch.

And then there was talk of a marching back past the police to the houses buried by the landslide. This idea chilled me again. I felt safe at the rally, but I did not know how I would protect my children from the police if they charged the crowd as they had on November 15. I didn't want to expose my children to this kind of fear. Frankly, I did not want them to totally abandon their respect for policemen as individuals and carry with them for the rest of their lives the knowledge that police can be ordered, unprovoked, to attack innocent people who are not breaking the law. I could see police cars positioned on the hill across the river, giving those officers a vantage point from above, in the path of my imaginary escape route. Suddenly, I urgently wanted to leave.

"Let's go," I said quietly.

"Go?" Suzanne asked, surprised. "Sure."

"Right now," I said.

Other people were getting the same idea, and we joined a pre-march back to our cars. We hastily loaded our things and drove out in a line of departing cars. As we did, we got a new perspective on the incredible number of police cars and officers. I headed north, turned around at the beginning of the Shively Road, and headed south again.

"What are you doing?" asked Suzanne, who, now that she is a teenager, likes to be home near her stereo and her telephone.

"I want to count policemen and cars. You and Rose count the ones on the left and David and I will count the ones on the right."

"Just on *this* side of the freeway?" she asked, getting her instructions straight.

"Yeah."

"Cars or policemen?"

"Cars."

We went down to the Pepperwood exit and came back, counting a second time.

"I couldn't get them all," Suzanne said.

So we did it again, this time counting at least a hundred cars.

Next we went up on Shively Road and watched the solemn march of people to the houses from above. I didn't know what I would do if it turned ugly, but I wanted to stand by. I heard yelling, and we left and drove once more past the marchers and the police. All appeared to be peaceful. Some of the policemen had taken off their riot masks. They would not be doing this if they were still ready to attack.

"Let's go home," I said.

"Yee-hah!" Suzanne exclaimed.

I heard later from friends that the marchers filled ceremonial sandbags and stacked them by the houses that are still threatened by the slide, and listened to speakers. Then the remains of the rally broke up and people went home.

Six thousand people had spent the day in harmony, quietly complying with the requests of the civilian peacekeepers. The police barely interacted with the crowd. They stayed out of sight on the freeway. Periodically they denied the public its right to free speech by preventing cars from arriving at the rally site. For hours we the taxpayers paid them to stand there, blocking the freeway, while nearby six thousand people took care of themselves.

Yet the newspaper reports infuriated me. "A mixed crowd, ranging from students, mothers with babies, longtime activists and elderly people in wheelchairs, defied occasional sprinkles and walked along a shaded lane to a large open field along the Eel River, lured there in large part by the appearances of singer Bonnie Raitt, Grateful Dead drummer Mickey Hart and former Governor Jerry Brown." What about fathers with babies? What about hale and hardy doctors and lawyers? What about a rising tide of people managing to inform themselves despite walls of press disinformation? "Lured?" What is the meaning of the use of the word "lured"? Lured by whom? Is the intention here to imply that the crowd was one of mindless incompetents with nothing much to do that day but be sucked into the woods by a few celebrities? Are we to ignore the fact that people deliberately rearrange their schedules and drive hundreds of miles to protest the destruction of our environment by big business, at considerable cost to their own peace of mind?

"Threats of felony arrests took the punch out of the much-heralded Headwaters Forest demonstration Sunday in the tiny town of Stafford." In fact, it was a nonviolent, nonconfrontational demonstration, as all of them have been except when the police have made them otherwise.

"[Darryl] Cherney told the crowd that Fisher Road was closed [to this year's rally] because officials didn't want the news media, and the world, to see a clearcut in that area." "A" clearcut? The fact is that if the rally site had not been moved to Stafford, satellite trucks parked along Highway 36 would have shown the world not "a" clearcut, but just how many mountainsides in one area can be stripped, burned, and sprayed with herbicides in one quick year with the blessings of our current system.

The story went on to say, "Along usually quiet Stafford Road, four generations of Thomsen family members who gathered at Mel Thomsen's home in honor of a son returning from basic military training were unhappy about the unexpected display down the road.

"Gregg Thomsen, 23, said he left a job at the Pacific Lumber mill earlier this year and enlisted in the Army. Married and the father of a 13-month-old son, Thomsen said he decided on the military 'because I know the timber jobs are going, thanks in large part to these people's environmental demands.'

"Grandfather Jens Thomsen said, 'It's a damn shame that this young man has to go away to find a future.' "

Douglas's iris

CHAPTER 45 MIKE O'NEAL

SUZANNE, David, Chuck, and I have just returned from talking with Mike O'Neal, whose house was in the path of the giant mudslide that flowed down the mountain last winter and into the community of Stafford. Only a single stand of trees just uphill from the house saved Mike and his sleeping family. Against those redwood trees, massive stumps and logs dammed and held back the torrent. As the standing trees became overwhelmed, they began snapping, instantly waking Mike and pulling him out into the rain to stand on his balcony and assess with horror the danger revealed by the dawn. At first Mike thought, "Gunshots!" Then he thought, "Bulldozers!" But when he saw seventy-foot-tall redwoods pushed over two and three at a time, he said, "The mountain!" Only a sudden breach in the log jam just a hundred yards to the side of his house saved him from the thirty-foot-high tide of mud that seconds later would have effortlessly buried Mike, his wife, his child, and their house.

Mike's own mountain, which he used to gaze up at from his rustic house, has now become his enemy. Once its slopes were anchored by huge redwoods, and a peaceful creek flowed into his garden and right under his house through a charming little culvert. Mike's ten-year-old daughter, Leah, used to play in that pristine creek, sending leaf boats into the darkness and running around to retrieve them as they emerged on the other side of the house. Last winter, when many of the tributaries in the Eel River system and the Eel itself had spread across their gravel-clogged beds, cresting their banks and continuing to spread into people's fields and homes, the creek that flows into Stafford mysteriously stopped. If Mike could have traced its course from its headwaters on the mountain, he would have seen that it had become clogged with debris and begun to flow down a dirt logging road, crossing the mountain face laterally, and saturating the unstable soil of the clearcut below.

Now, almost a year later, the very trees and earth and water that were once the majestic backdrop of Mike's life stand poised to move again. It is autumn, and the entire slide that lies hardened and inert needs only rain to set it in motion. The little creek has lost its innocence. Part of it is now subterranean, flowing somewhere beneath the mass of soil, while part of the creek has managed to surface in the wasteland, flowing again with the grace of any little rivulet, as if trying to return to normalcy. But one stares down at the water and up at the mountain and over at the fifteen-foot-diameter old-growth redwood stumps lodged where they tumbled high onto rough walls of gravel, and it takes almost no effort to start the projector and run the movie past where it paused last winter, mid-tragedy.

When I cut the tips of my fingers off last spring, what struck me was how there is no musical buildup to accidents in real life. I remembered this as I walked the slide with Mike. No music plays to the aftermath of tragedy either.

But another element is missing. At the scene of real-life tragedies one usually finds an overkill of emergency vehicles pulling in with flashing lights, buffering the rawness of reality with the appearance of efficiency. Later, social agencies lend their support and consolation. Since Stafford was buried, however, little of this has transpired. Virtually no one has responded to these people's plight. In a county where the media and many of the agencies that are supposed to serve the public appear to be held in the palm of the timber industry, the eerie scene is one of helpless people and dried mud. Even though several of the residents are retired PL employees and others are truckers who have worked for the company, MAXXAM, which in 1991 logged the nearly vertical slope where the slide originated, and returned again in 1993 to clear-cut another area on the same slope, returned after the slide occurred, not to help the people but to pull its own logging equipment to safe ground.

The residents themselves, trapped by the deep mud, were using bolt cutters to make an escape route through a chain-link fence that divided them from the adjacent freeway ramp (the same one, by the way, which was recently lined with orange traffic cones and squad cars), when suddenly a Cal Trans highway worker showed up, not to help, but to aggressively ask who had given them permission to cut the fence. Members of the National Guard appeared a few days later, reassigned from the crisis in Ferndale, where Francis Creek, which was also running wild with logging debris, was joining with the Eel to flood the valley's dairies and homes. The National Guard cleared the main roads in Stafford, and a few days later MAXXAM returned, again not to lend a hand to the residents, but to extract the valuable logs that lay every which way amid the mud-filled houses, the flattened trailers, the demolished vehicles, the dislodged fuel tanks, the remains of outbuildings and the scatterings of belongings—dishes, furniture, clothes, family pictures, children's toys—that had lifted with the surge and come to rest at random.

A few days after this year's Headwaters Rally, after puzzled reporters photographed the houses and questioned why they were being preserved in their graves, MAXXAM, who has bought out those landowners who will settle, showed up to "erase" the houses, terrace the soil, and plant a lawn. As one friend of mine said, "It looks like they are getting ready to build 'Stafford Estates.'" But the sad fact is, the land can't be resold. Because of the threat of more debris torrents, residents like Mike, whose houses are not yet destroyed, can no longer live in them.

"Are they going to buy you out?" I asked Mike.

"No. Because my house didn't actually get flattened, but still can't be occupied, they offered me twelve hundred dollars an acre, same as they do for timberland. I've got one-third of an acre, so that's four hundred dollars. They told me if I want to get paid for my house I should sue the people I bought it from, saying it was built beneath an 'old slide.'"

"It looks like there were ninety- to one-hundred-year-old trees growing on that 'old slide,'" Chuck observed.

"Yeah. The mountain hasn't moved since 1901. If it was still so 'unstable,' what were they doin' loggin' on it? Why'd CDF give them permission? My house was built fifty-seven years ago. The county gave permits for the Johnson house to be built 'cause, with timber on it, that mountain wouldn't have gone anywhere.

"But two years after they clear-cut it in '93, I told them something was wrong. The gravel was comin' down and fillin' in the creek channel. I was gettin' flooding and gravel around my house. I didn't get any help, so I finally went myself and put a piece of plywood across the culvert on the other side of the road so the creek couldn't come under my house anymore. But PL sent the sheriff's department out to remove it."

Mike's house was charming. Even though he lives in a community that was once "supported by timber dollars," Mike, like many of his neighbors, has a deep love of his natural surroundings. He has built balconies and decks all around his house.

"I used to be able to see down to the Avenue of the Giants in Humboldt Redwoods State Park from up there," he said, looking up at his main deck. "I used to take my telescope out there to look at the stars."

In Mike's small house that he rents now in the logging community of Rio Dell, across the river from Scotia, the biggest personal telescope I've ever seen occupies a prominent position in the living room. From the first time I spoke with Mike and his wife, Joanne, I was struck by their global perspective. Like Kristi, and many other residents of various watersheds who have suffered loss of private property and peace of mind, Mike seems strangely perfect to be one of the people who have no choice but to stand up to MAXXAM. He can't sell his house because he would have to disclose the danger it is in. He can't rent it out for the same reason. His insurance company has canceled his house insurance. He doesn't have the money to buy a new house and walk away from the situation. Right now his house stands abandoned, simply a liability, while Mike is forced to rent at his own expense in Rio Dell.

"This is a beautiful barbecue," I commented, admiring a circular structure of extremely precise brickwork.

"I built that last fall. Never got to use it. Now people are using it for trash. They stole my firewood . . ." He scanned the yard.

"I planted those fruit trees. I had a big garden over there," Mike pointed out as we strolled up his driveway and onto the road. We came to the culvert where the muddy piece of plywood still lay on its side where the sheriff must have dropped it after pulling it free, and walked up the hill of soil and gravel that covers the road.

"There used to be a downhill slope here. My daughter used to *coast* down here on her bicycle." We stood staring up an incline, trying to perceive the volume of the debris that had hardened in place before us. Mike struggles to try to convey the immensity of what he experienced. I have seen him over and over describe his terror, looking up again in his imagination at the wall of mud and water that was coming his way in the first light of December 31, 1996.

We walked up onto the slide, and I looked up the mountain over a vast wasteland. Though I had seen video footage of the devastated area, as often happens, the camera could not capture the effect of the wide swath of destruction.

"Oh my God," was all I could say as I worked to make sense of the scene.

"These are the trees that held it all back. These trees . . . ," Mike said.

He spoke with obvious appreciation for the trees that had saved his life, as if they were heroes. By redwood standards, the little stand of approximately ten 70-foot-tall redwoods were relatively small. Piled up behind them were eighteen feet of mud and a pile of logs, many of which were bigger around than the standing trees themselves.

"That log must be eight feet in diameter," Chuck said, measuring with his eyes. "It must be fourteen feet long! I bet that log alone is worth four to five thousand dollars!" And then he stopped, spotting for himself what Mike was about to put into words.

"Look at the bases of the standing trees," Mike said.

"I see," Chuck said.

My eyes focused past the gigantic mound of debris to the trunks of the trees that held it. I don't think anything this whole year has hit me with the same force as the sight of those trees. The very trees that had saved the lives of Mike and his family were marked for salvage logging. We scanned up the flow, and all along the edge of it the trees were encircled with the same blue lines that indicate "Cut."

"These trees are the riparian corridor that they were required to leave when they clear-cut!" Chuck said. "Now they think they can get a salvage-logging exemption and take them! This land belongs to Barnum Timber Company, but they're working hand in hand with MAXXAM. They've *caused* this catastrophe, and now they're going to make money on it!"

"Yep, they're two dirty peas in a pod," Mike agreed.

We looked upslope, and we could see light through the trees. Only a narrow corridor of trees remained.

Mike set off walking up the flow, pointing at the wounds twenty-five feet high on the trunks where the trees had been hit by floating debris before the log jam broke. These wounds, combined with the fact that the trees' bases stood in hardened mud, were apparently what qualified the trees to be salvage-logged in the eyes of the forester with the blue paint. In fact, the trees were perfectly straight, tall, and healthy. Redwoods are not made vulnerable by damage to their bark the way other trees are. In the life of a redwood, these skinned spots were but incidental. For me, however, the whole scene was final proof that the indignation I have been feeling is not out of proportion to its cause.

I pointed up at the mountain. "Mike, is MAXXAM still going to log adjacent to the clearcut that caused this slide?"

"Yep, they may be starting on it this week."

"I don't get it," I said. "How can CDF have approved that plan?"

"They approved it exactly one month after this slide occurred," Mike said.

In fact, they approve almost all plans. If they don't like a plan, it is sent back for mitigation, that is, for a few changes to be made, and then it is approved. They almost never outright deny a plan.

I asked Mike, "What happened after all this came down?"

"Well, it came in three surges. The first surge was at seven o'clock A.M., when I got everyone out of their houses.

"The second surge surrounded a mobile home, and then the mud began creeping into people's backyards."

"It was creeping?" I asked, somewhat hopefully.

"Well, between surges. But the surges were in monster blocks, eighteen feet high, coming as fast as you could run.

"The third surge was the next morning, New Year's, 1997."

"*Twenty-four hours* after the first?" Someone was slowing down the camera, and I was beginning to get a grim picture of just how alone these people had been. "You had *twenty-four hours'* warning and still no one did anything?"

"It was really strange. After the first surge, I called 911, and a sheriff's deputy came by and said, 'This looks unstable. You people better leave,' and then he left and never came back. The Office of Emergency Services said they might be able to help us in a couple of weeks. We had to go *find* the Red Cross, and they gave us a couple of weeks in a motel room. FEMA, the Federal Emergency Management Agency, finally came out after the mud had solidified and gave me eight hundred dollars for my foundation blocks, but they lost my paperwork three times. The county and the state didn't want to deal with us because of 'the responsibility.' "

The responsibility. This word gave me pause. What is really at issue is "liability." I learned from a National Guardsman that a county sheriff's helicopter actually flew over the impending slide twice, first to shoot videos of the cracks on the mountain face and later to film the damage. When members of the County Board of Supervisors were confronted, however, and asked, "What have you done to help the people of Stafford?" one of them was brave enough to honestly confess, "Nothing."

I thought of the hundreds of officers who stood shoulder to shoulder, arms folded, during the rally.

"There were *no* policemen?" I double-checked incredulously.

"No, I never saw one policeman," Mike repeated, politely.

"What about the possibility of looting?"

"There was looting after we left." Then he paused. "I don't think you quite realize the degree of callousness I've been dealing with," Mike said, with touching understatement.

But he had saved the most chilling news for last. He said it rather quietly, as we were walking back toward our car in the waning light.

"They want to run the creek back to its original course under my house."

"What?"

"At a recent meeting, MAXXAM told residents that they're going to run the creek back to its original channel right through my culvert."

By then we were passing beside his house. I looked down at the culvert, now half filled with sand, with only twenty inches of clearance remaining. It was hard to believe that a creek that could fit through such a small hole could do so much damage.

"I think they want to flood me out 'cause I won't be quiet," Mike said.

I thought of all the other residents in the other watersheds who feel the same way. They just want to enjoy the peace and quiet of their daily lives. Mike just wanted to enjoy his balconies and decks, his view, his telescope, his gardens, with his wife and Leah. These are such modest, balanced desires. Yet greed, which by nature is unbalanced, attempts to steady itself by becoming parasitic on the lives of others, and in the process innocent families, communities, and whole ecosystems become unbalanced. This is the unholiness of what is happening.

CHAPTER 46 THE DIPPER

JUST when I thought I couldn't read one more newspaper about Headwaters without losing my mind, I yielded to David's requests to go have some fun and took him and a new friend of his, Andy, out to a campground east of Arcata on the south fork of the Trinity River. I discovered the place a while ago and planned to return to write while David and a friend played in the creek. I parked my camper, the boys climbed out, and soon I heard that delicious crashing of rocks moved by ten-year-old boys damming water. As I listened to the rocks my muscles began to relax, crash after crash after crash.

This is a classic sound of childhood. It is like some sort of psychic acupuncture. Each boulder that lands hits a nerve and releases pressure. Thank God ... Thank Gaia. I needed to know that there are still boulders in creeks ... and cobbles. I needed children to come back to me cupping tree frogs in their hands for a momentary peek. I needed to see fingerling trout lingering in a pool, unaware of just how rare they and their pool are in this beleaguered coastal forest. I needed to see the dance of reflections on the leaves of alders overhanging the water. I needed clarity, and here it is.

We spent the night there. Chuck joined us and then left for work in the morning. The entire campground was empty except for David and Andy and me. They wanted to show me their dam, and then we meandered upstream through ancient Douglas firs, farther and farther, drawn beyond the last campsite to a wild stream course that belonged to a single dipper eating insects.

"Wait. There's a dipper!" I exclaimed. David has watched lots of dippers, but Andy has only recently become our friend. I did not know whether he knew about dippers. So I paused, waiting to see if he would ask in his polite way, "What is a dipper?"

"What is a dipper?" Andy asked.

"See that little bird up on that rock?"

"Where?"

"On that flat beige rock on the left shore of the creek just below that gray rock with the three lines in it that go up . . ."

"Yeah!"

"That's a dipper."

"It went in the water!" Andy exclaimed.

Dipper

"That's what dippers do. Watch it," I said.

"Oh my gosh, it went under the water. Oh my gosh!"

The dipper had no concern for our presence. It was exhibiting classic dipper behavior. It walked across a rock, down the side, into the water, disappeared beneath the water, surfaced, beached momentarily on a tiny shore, turned, reentered the current, and disappeared under water again as the current carried it into the next pool, where it calmly emerged on shore, bobbed up and down, turned, and walked back into the stream.

"What is it doing?" Andy asked, as if he were watching a bird that had gone insane.

"It's hunting for insects. It walks along the bottom picking up aquatic insects from between the rocks. Its feathers are coated with oil. Have you ever looked at a chicken that your mother was going to roast? Have you ever seen that thing above the tail, that gland with a little hole in it?"

Andy politely said, "Yes," though he sounded a little tentative.

"That's the chicken's uropygial gland."

"Oh," he replied politely again, not daring to ask what a uropygial gland is.

"Birds take oil from their uropygial glands when they are preening their feathers," I explained, "and spread that oil all over their feathers to make themselves waterproof. A dipper, even though it is so small, has a uropygial gland as big as a duck's!"

"Oh my gosh," Andy said, staring at the dipper.

"Count how many seconds a dipper can hold its breath," I suggested, and then we stood in silence simply watching.

"Forty-five," Andy said.

"Forty-five what?" I asked, already forgetting the question.

"Forty-five seconds." This seemed a little long, but I was glad Andy was impressed with this unusual little bird's capacity to stay under water.

"They often nest behind waterfalls," I said. "They build a nest in a hollowed-out ball of moss that actually stays wet from the spray of the water. The female climbs inside and just sticks her head out of the hole to get food from her mate." I spontaneously did my female-dipper-sticking-her-head-out-of-the-nest imitation, a mere forward extension of my neck.

"Wow," Andy said, politely.

We watched the dipper a little longer.

"It's feet aren't even webbed!" I said, underscoring its difference from waterfowl. I can't help but seize any opportunity to inform anyone at all about the natural world. It is therapy for me. The clarity of the water, perfect habitat for the dipper, which uses transparent eyelids like goggles beneath the water to spot the insects, cleared my mind. I cannot see another silted creek for a while. I cannot endure any more absence of dippers. Here there is no mud. When we walk through the water, clouds of silt do not kick up behind us. I feel as if the forest has delivered me just in time to an environment that refuels my soul. This creek flows out of pure old-growth Douglas fir.

> *I come to a place like this creek, which is in a state of holy equilibrium, and it is* enough*—granite and quartz and giant leaves which my son gathers as specimens of the giantism induced by life in the shade, this cathedral of alders, catching the light far above in the sap-green panes of thousands and thousands of leaves. The dipper is the minister, stepping up to his pulpit, a small slab of granite, pausing as if to address us all, and then slipping onward, methodically down the rock into the water. He reappears, pauses as if he might address us still, and then turns and submerges again. As the current carries him through a tiny riffle and he gets a foothold and emerges, I*

ask myself how I could need any more than this. It is so fascinating. Diversity. Biodiversity. The opposite is an aching, insane boredom that is enveloping our children and our marriages. We think the other species suffer from the daily extinctions on our planet, but humanity also suffers.

People speak to me about the importance of teaching the next generation to love the natural world so that it will be safer for all of us. Sometimes I am reluctant to teach my children to appreciate a bird like the dipper, because I feel that I am setting them up for loss if the dipper's habitat is destroyed. I have secretly begun to suspect that we are unfairly asking children to do the work that rightfully falls to those of us who are currently adults. I think it is hypocritical to soothe ourselves with the idea that we are doing our part if we are *only* teaching the next generation to do theirs.

At night, the boys and I made a campfire. I pulled a few pages of newspaper out to read instead of crumpling them. Two articles, side by side in the section of the *San Francisco Chronicle* called "Nation," were titled "Report Shows Rich Keep Getting Richer" and "Preteens Say Drug Use of Peers on Rise." The former informed us that the wealth of "the nation's richest 2.6 million, the top one percent, was about equal to that of the 88 million people in the bottom 35 percent of income." My eyes moved to the article on preteens. "The percentage of 12-year-olds who say classmates or friends use illegal drugs more than doubled in the past year. . . ." Just as the lack of woody debris in our creeks is causing our watercourses to scour, erasing pools, eliminating diversity, so stripping our planet of life is scouring the souls of our children.

I stood watching the dipper with David and Andy, and I thought, "This dipper needs just a small quantity of bugs for this day to survive. To need more would be grotesque. It needs simply a day's worth. Tomorrow, another day's worth. In this holy cathedral this is the sermon, day after day the same."

On the way home we stopped by to visit friends who live in the mountains, and after introductions were made, the first thing Andy told them was, "We saw a dipper."

CHAPTER 47 LUMBER

DOWN the road from me, a new couple in the neighborhood have cut a hole in the east wall of their house, and a week ago the pressure-treated underpinnings for a new deck were constructed. Today, as I drove my children to school, I saw that the deck itself has been built of old-growth redwood. I rounded the bend and there it was, that beautiful red color of the heartwood. It is a wonderful material. All of the houses around here are built of it.

Yet, because of the way that the last of this old-growth redwood is being acquired—carelessly, with no thought for the future—it has been *made* to be an inappropriate material. Instead of this wood being squandered, it could have been cut sustainably and priced in proportion to its outstanding qualities as a building material. For special projects like a deck, that will be the focal point of countless people's lives for the next hundred years, it could *still* be the perfect choice. But it isn't now.

Old-growth redwood is usually sold with the wood grade designation called "Clear Heart." In a ban similar to that imposed on the purchase of ivory or rhinoceros horns, many retail lumber firms have agreed not to sell this product. If you do see it for sale, contact the management of that supplier and tell them that you buy lumber where *certified sustainably harvested* redwood is sold. Not only will this help control the market for old-growth redwood, but it will help support timber companies that are using sound forestry practices. In combination with limited use of sustainably harvested wood, consider using composite lumber made out of recycled plastics and wood. It is widely available, and, like redwood, it won't rot and doesn't need sealing.

In my house, I want to cut a hole in my own east wall and put in a door with a deck. A couple of years ago a friend rebuilt a whole studio for me out of old-growth redwood recycled from a grocery store that was weakened in an earthquake. Most of the boards simply needed to be swept off before they were cut to size. Others that showed signs of rot he shaved on a table saw, and the wood looked like new. Every year buildings are torn down and the wood wasted. Scaling down our projects and using salvage materials wisely are two other solutions. But I think I will look into plastic "wood" made from recycled containers.

Recently I stopped by to see an old friend who, ironically, works at a lumberyard that has an entire annex devoted to redwood. I had never seen so much redwood lumber under one roof. Alongside wide, fine-grained red planks of ancient heartwood, I was startled to see narrow, yellowish boards of wide-grained sapwood. It is more and more common to see trucks loaded with young, spindly, ten-inch-diameter redwood logs, but I had never seen those logs after they have been milled into boards. No sooner did I notice the boards than I instinctively averted my eyes. I felt as if I had walked into a factory that exploited children. I felt the same quandary that one often feels on the road these days. It is hard to decide if it is more sickening to see a truck loaded with one or two gigantic old-growth logs, or logs that have been taken before they ever had a chance to mature.

CHAPTER 48 ALTERNATIVES

A FEW weeks ago I visited the woodlot of a rancher named Rudi Langlois who has become outspoken about MAXXAM's logging of the Freshwater Creek watershed. The Freshwater area was clear-cut early in PL's history, before A. S. Murphy took over and instituted his policy of selective cutting. It was then ignored by the company until recently. I remember when Chuck came to my house in January of 1997 with a new map, just released by PL, entitled, "First Decade Harvest." The name refers to the first decade of their now obsolete Sustained Yield Plan. With regard to Freshwater, the map is still all too relevant. It shows PL/MAXXAM's entire holdings in white, with large patches of green, and accents of pink and purple. It is a cheery little map until one looks at the key and sees that green equals "clear-cut." The two largest green areas by far are in the Freshwater Creek drainage and the pristine area surrounding Grizzly Creek Redwoods State Park, which is past Carlotta on Highway 36. When this map was released, a collective groan was heard from the environmental community.

Rudi Langlois and his neighbors are living the reality of this map's graphics. Rudi's children, now grown, played and fished in Freshwater Creek. We stood on a bridge and looked down into the dirty water, however, and this was unimaginable.

"My friends ask me, 'What are you doing, Rudi, becoming one of those Earth First!ers?" I don't care anymore. I have to speak out about this."

I had come to see Rudi's forest, a large tract of which he is enormously proud. He has logged it several times, once just a year before my visit, yet there are large trees; the understory is lush; there were no skid trails; and birds sing throughout the forest.

"You wouldn't believe the growth on those trees after we took the larger trees out." Rudi's voice relaxes and expands when he speaks of his own forest and tenses when he speaks about the clear-cutting that is enveloping his valley.

Recently I received my copy of the long-awaited *Headwaters Forest Stewardship Plan,* which is subtitled *A Citizen's Alternative to MAXXAM Management of Headwaters Forest.* It has been prepared by the Trees Foundation in Garberville, California, in collaboration with the Institute for Sustainable Forestry and a coalition of other environmental organizations. It is an extremely technical report, filled with numerous tables and graphs with headings such as "Projected Size and Density Class Development," "Projected Restoration Forestry Harvest Volumes," "Net Revenue from Headwaters Forest Under Certified Forest Management," which requires extra concentration from those who are not Registered Professional Foresters. But a quick scan of the table of contents, the photographs, and the maps simply gives one a solid sense of hope.

The opening page reads:

> *The Headwaters Forest Stewardship Plan (HFSP), born from a deep concern for the integrity of the redwood ecosystem and the viability of timber-*

related jobs and revenue, offers an alternative for management and land use of the 60,000-acre Headwaters Forest. It is presented to the community of Humboldt County and concerned persons elsewhere, including the timber workers, conservation community and government agencies.

Based on accepted scientific principles of conservation biology, the Headwaters Forest Stewardship Plan puts forth a three-pronged proposal for preserving the existing ancient redwood forest. This is achieved by setting aside pristine core reserves, restoring cutover lands to suitable mature-forest habitat, and carrying out responsible, long-term forestry in appropriate areas of second-growth forest. Economic analysis clearly reveals that significant levels of employment and revenue will still be generated from Headwaters Forest, through implementing forestry prescriptions and restoration activities.

I can think of no more worthy critics of MAXXAM's liquidation logging than those people who practice forestry as an art in the middle of tragedy.

CHAPTER 49 TOXICS

CHUCK and I met for coffee at our local cafe this morning. Ahead of us in line was Patty Clary, director of Californians Against Toxics. I asked her what was new. Her eyes came to life with the news that the County Health Officer of Mendocino County had written a letter to the County Board of Supervisors stating that spraying herbicides on clearcuts is a danger to public health. This may not sound very impressive, but it is just another example of the fact that it takes forever to get paid officials to say what knowledgeable volunteers have been saying for years.

Our conversation wandered to the role of diesel in herbicide spraying.

"What I want to know is, what about the diesel fuel they use as a vehicle for the herbicide? Does anyone keep track of that?" Chuck asked.

"No," was Patty's flat response. "MAXXAM sprayed two thousand gallons of Atrazine last year. They have to report that, but they don't have to tell us how much diesel was used as a vehicle for that Atrazine or the other herbicides they spray."

Then she drew a little closer to us and said, "I was talking to a licensed pesticide applicator the other day who said that PL hauled a six-thousand-gallon tanker filled with diesel fuel up the Shively Road and *used all six thousand gallons on 320 acres.*"

"But are there any limitations on diesel use?" Chuck asked.

"I don't know. I need to investigate this. Diesel is far more persistent in the environment than herbicides."

"If that same six thousand gallons were spilled on the highway in a wreck, the government would be all over the guilty party to pay for the cleanup," Chuck continued.

"That's right," Patty said. "But timber companies do not even have to notify that they have sprayed herbicides and diesel, except in a monthly report made at the end of the month *after spraying has occurred,* and only the general location is indicated on the map. The fact that licensed applicators are doing the spraying is supposed to ensure public and environmental safety."

As I listened to Patty, I remembered her file cabinets like cauldrons, filled with information on poisons. In the seventies she and her children were victims of aerial spraying during the medfly scare. She has been working on the issue of toxics ever since. But this task should not fall on the shoulders of one victimized person. For her own health, her burden should be shared by others.

CHAPTER 50 A WALK ON THE BEACH

I HAVE been sick with a cold for the past week, but I finally feel well enough to go for a walk by myself on the beach. It is a long, cliffless, rockless beach that rises into low dunes unnaturally secured with introduced dune grass. As I walk along, I let my mind wander, secure that the storm waves and an unusually plentiful number of sanderlings and sandpipers working the waterline will ultimately lift me out of the humorless slump into which I slid with my first symptoms.

In spite of all the wonderful people I have met, and in spite of the fact that I would commit to this book again without a moment's hesitation, the subject of Headwaters gets grimmer by the day. At the same time, it also gets more and more fascinating as it convolutes and grows nearly beyond my comprehension in complexity. In its most complimentary light, it takes on the appearance of a rich, well-crafted novel that one leans forward to watch move deliciously toward resolution. In its flattest, harshest light, it reminds me of the Frost poem, which wonders whether the world will end by fire or ice, only the question in my mind is, "Will it end with people losing their way in piles of paperwork while machines work with cruel efficiency to strip the Earth of all that sustains its creatures?"

Recently I spoke on the telephone with a professional forester from Santa Cruz. I called him because word reached me that he had been on a tour of Yager Creek last year, hosted by MAXXAM, and that he was so stunned by the condition of the land around him that he had to pull away from the group to get his bearings. In our phone interview, he said that he could not believe that MAXXAM officials would take foresters and other agency professionals on that particular route. He said he thought they would be ashamed of the sins around them, rather than holding that piece of land up as exemplary of their operations. What confused him even more was that only one other professional seemed appalled by what they were being shown. Most of the men were passively nodding and gazing, fully accepting the words that were being spoken and ignoring any little, murmuring voices of professors past who might have taught them genuine forestry.

In technical terms this man ran down a list of crimes against the land, as if he had seen it yesterday, that exactly matched what I feel I have observed in Owl Creek, Freshwater, Yager, below Headwaters on the Elk, in Bear Creek, on Strong's Creek . . . This professional confirmation made me feel momentarily uplifted and relieved. I wasn't exaggerating. I wasn't overly influenced by the passions of other people. We are talking objective reality. The mud that almost killed Mike O'Neal is real. But then this confirmation of truth settled with my cold and weighed me down as if I were one of the houses in Stafford.

I have come to this beach to be renewed. I walk south parallel with the town of Samoa, and I can see the Louisiana Pacific pulp mill up ahead putting out steam. I think of Baja, of the warmth and the birds, and I long for honest rock, naturally devoid of trees. I long to simply paint, to escape controversy, to escape reality. I want color to speak only of itself, and form to speak only of itself. I want the realm of my imagination closed, so that I am the last word. And yet, I know too much for this. I have lost this innocence. I have learned that my life, disconnected from the fight to preserve what I love, is too uncomfortable. It was only when I linked beauty with reality that my life really took hold.

I return to town and take a sauna. I rarely treat myself to a sauna alone. The silence and warmth hold me. I listen to the whisper of the heater beneath the rocks and drink a hot lemonade. I put eucalyptus oil on the rocks and my breathing clears.

I stop by the store to buy dinner. On the newsstand I see announced in the San Francisco Chronicle *that the purchase of Headwaters may be put on California's June ballot. The novel convolutes once more. I knew that the issue was scheduled to be put before the public, but seeing it in print fills me with apprehension. Will the forest's truth win out?*

The issue will be illuminated by whatever light the maker of each television or newspaper ad chooses to shine. Ironically, MAXXAM, *which is in favor of the Deal, will be scrambling for the best shots of ancient trees, spotted owls, and marbled murrelets, telling the citizens of California about the wonder of the forest; while the environmental community will be showing photographs of smoldering clearcuts, debris torrents, falling old-growth trees, trying to pull the sheep's mask off the wolf. From their couches, Californians will not feel the mud, nor peer through pampas grass across clearcuts to the few isolated stands of forest that are the subject of the controversy.*

But perhaps the forest's truth will prevail. Scientists have learned the barest rudiments of the language of the soil, of the mycorrhizal fungi, the red-backed vole, the flying squirrel. They know that we must speak not only about lumber and trees and jobs, not only about the animals that live in the canopy and on the trunks of the trees, but about the downed logs, and the soil-making process, and of the soil itself. We must speak of water and the creatures of the water, of the lost souls of the salmon and the livelihoods of those who fished them. Regardless of the Deal, and the outcome of this vote, we must use this opportunity to let Gaia speak her truth and to educate the public about this type of corporate greed that undermines the carefully crafted laws of the people to steal the forest from its heirs.

Huckleberry

CHAPTER 51 PERSPECTIVE

WHEN I spoke to the forester from Santa Cruz, I asked him to send me a follow-up letter about the logging "show" near Yager Creek. The information had come so fast and was so technical that I was unable to take it down over the phone. He promptly obliged. I hold in my hand a document that assures me that I am not caught up in some hysterical vendetta against an innocent corporation.

His two-page, single-spaced letter revealed the following observations:

The project site was surrounded by Timber Harvest on over 50 percent slopes that lead straight to the creek. This THP completely disturbed all the ground, mixing all the soils, "A" horizon with "D" horizon soils and no ground was unturned. I asked a company employee standing by me, "Why did you use D-8 grappler cats on this steep ground, and why so much disturbance?" He said that he had also questioned this approach to log removal, and was told, "We have the equipment here, so that is how it's gonna be done...."

The cats had driven blade-down over all the ground and taken a different path to each log making no ground safe from the blade work. The disturbance was so severe as to kill redwood stumps... no undisturbed buffer was left along the creek.

The general site conditions were very surprising: killed redwood stumps with no sprouts, huge invasions of non-native plants (pampas grass, broom) and hundreds of acres with ceanothus and invasive softwood and little else. Ambient air temperature soared in the area while they blamed the farmer for delivering 70 degree water from the top of the ridge. This was always true in the time before logging in this basin, except then the forest cooled the water down to 55 or below. This was all occurring while up on the gentle slope was a Skyline Yarder working ground that was more suitable for cat operations, a Skyline Yarder that should have been employed on the 50%+ ground that had been logged by the D-8 grappler cats.

The final insult was to have steelhead smolts drained out of a hatchery rearing pond into the creek for show. This may seem innocuous at first but you must consider that (1) steelhead are probably the last fish that need to be propagated. Better yet would have been coho or chinook; (2) the timing of release must be precise to insure that the release is timed to coincide with lunar phases and smoltification that lead fish out of the watershed and to (3) reduce competition with existing wild fish in stream by insuring reduced residency of hatchery fish.

All a very depressing sight and a good show as to how deluded and unaware these landowners are of the destruction they lay on their watersheds.

Having seen the miles of destruction, it is a relief to have other people begin to acknowledge the same reality that Doug has shown me. An occasional agency employee stands his ground and states that the emperor has no clothes, but frequently he is then transfered, uprooting his family if he has one. Gradually, our newspaper, pressured by the courage of the local people and an increasing number of letters to the editor, is beginning to report a version of events that I recognize. Each time someone tells the truth, it makes it easier for the next person. Conversely, each time someone looks away, it undermines us all.

CHAPTER 52 HIGH PRESSURE

DOUG is insisting that I see all six groves of Headwaters. I have not yet seen All Species, Shaw Creek, and Elk Head Springs. It is so easy to see the forest as an abstract symbol of loss rather than realizing that these groves are real places, that they stand out there day and night, that each is different, that most of the animals and plants that live in them know no other place but their respective groves.

Many people working on this issue have never seen any of the groves. They have only heard the statistics, read the conditions of the takeover, and computed right and wrong with their own souls. Or, perhaps they have seen the groves and the surrounding wasteland of clearcuts from the air. They may have seen the flooding, the contamination of water downstream, the clearcuts sweeping reality around to the front sides of the mountains. Or perhaps they have seen Doug's slide show and heard his compelling description of all he has observed over the past six years.

Until I came along, Doug was the only person, as far as we know, aside from MAXXAM employees, who had seen all six groves. One by one, however, they are becoming a reality for me, a little against my will. If I got a break from parenting, it would be nice to spend it relaxing in the sun, swimming, or hiking in an ecosystem that is still at peace with the twentieth century. But Doug is insistent, and we have picked a day to go to All Species and Shaw Creek in one trip. I get out my pack, inventory its contents—a fleece, rain jacket, rain pants, flashlight, lightweight long underwear, bandana, shorts, lightweight hooded sweatshirt, wool gloves, wool hat, extra socks, water bottle—and listen to my weather radio, thankful that a high-pressure area is stationary off Cape Mendocino.

CHAPTER 53 ALL SPECIES GROVE

"SEE those trees in the distance? That's All Species Grove," someone pointed out to me months ago from a ridge in Kneeland, east of Arcata. The isolated stand of old-growth redwood and Douglas fir looked small, but intriguing.

A few weeks ago I drove up on the same ridge with my children, trying to track down some Manx kittens that had been advertised in the classified section of the newspaper. I gazed over to lay eyes once again on All Species and was surprised to see that it looked like a ragged desert island in the middle of a sea of clearcuts. Was this the grove that had intrigued me six months ago? It looked more like an eyesore. Had I mistaken its actual location? I scanned right and left across the skyline, looking for a thick stand of old-growth. My eyes came back to the same spindly trees, tall and thin because they are adapted to life in a dense forest, looking more like solitary palm trees than redwoods and Douglas fir. I had intended to point out All Species to David and Suzanne, but I felt embarrassed to admit that I was stealing time from their lives to try to save such a grotesque fragment.

"Did something happen to All Species Grove in the past few months?" I hesitantly asked a Doug, feeling a little silly about the awkwardness of the question. It can't be logged until the Deal is agreed upon. I knew it could be salvage-logged, but when I think of salvage logging, I mourn for the forest floor, not the canopy.

"Until recently there was a buffer of residual forest around All Species. Now that's gone. And it's been heavily salvage-logged in the past few months. That means 'diseased, dead, or dying.' More of the grove drops down over the ridge, though, on the other side of the slope."

Salvage logging—free rein to have their way not only with downed logs, but with whatever they decide is "dead or dying." I was suddenly anxious to go to All Species Grove to erase the image of palm trees from my mind.

"I have a spare hour. I'll hike with you for a while," said Chuck, who had agreed to drop Doug and me off.

As we hiked down through oaks and fragrant California bay trees, I pulled off a leaf of bay as I often do at the beginning of a hike, crushed it, and inhaled its pungent scent to clear my sinuses and say hello to the forest. Surprisingly soon, we hit an abundant stand of ancient Douglas fir. The trunks were only a few feet in diameter and seemed understated due to their relatively small girth and the lack of lichen, but they towered high above us, dry and serene.

"Is this All Species?" Sometimes I have to ask. What little remains of the old PL's thinned groves often looks like virgin forest. There are no signs that say, "All Species Grove." For MAXXAM, all the forest that is left standing is viewed simply as a resource which has not yet been fully exploited, like gold in veins that has not yet been mined. The company does not readily acknowledge "groves." Instead, once it has targeted areas for cutting, it speaks only in terms of Timber Harvest Plans. The groves have been identified by environmentalists, after they have become isolated by clear-cutting from the larger forest that once washed across the valleys like a sea. But, of course, I had only to look down to see that there were no stumps. This is one of the wonders of an old-growth forest. There are no stumps. Nothing has been extracted except by nature.

Each grove has a story that accounts for why it is still standing. By now they would all be gone, without intervention.

"How did All Species wind up being saved?" I asked Chuck. "How come it's still here?"

We had reached the far end of the grove and entered a clearcut. Chuck was at the opposite end of a long redwood log that the three of us used as a raised path through the rough slash. Chuck turned, and in a rare flash of pride said, "Because when it was scheduled for cut, I researched the grove and found there was a case here."

"Murrelets?" I asked.

"Yes."

"You mean, without *you* this grove wouldn't be here?" I asked incredulously.

"I guess not," Chuck said simply, but with a smile that admitted that he was pleased to be asked.

"If I hadn't asked, you wouldn't have told us that, would you?" I pressed.

"Probably not." And this is so. Chuck reminded me of the Douglas firs upslope with their quiet, unspoken power.

As we hiked together toward Bell Creek, the map continued to come to life, to gain dimension. All Species now had trees, and Bell Creek had changed from a silent blue line to a brisk little creek passing beneath a metal bridge. The intriguing sound of a small waterfall came from the direction of some mossy boulders upstream. Doug and Chuck dropped down an embankment to explore the creek. I followed, reluctantly, wishing instead to save steps and time for the long hike ahead. But the minute I was by the water, I just wanted to head upstream, instantly seduced by color. Because it was autumn, the leaves of the maples were a tender yellow, like the delicate skin of the elderly. The pool above the falls was a deep steel gray, and above it hung brilliant red huckleberries that looked like the dyed salmon eggs that come out of a jar.

"Salmon eggs aren't *really* this color, are they?" I asked Chuck.

"No, they're orange. I hate to confess that when there used to be salmon in the Van Duzen, we used to catch the females and use their roe for bait."

Doug was silent.

"Just another bit of information out of my past," Chuck said. Chuck is fifteen years older than Doug. He and I were raised during a significantly less enlightened and more abundant era.

"I better get back if I'm going to get David to soccer practice," Chuck said. He had agreed to take care of my children so that I would be free to go to Shaw Creek. I could see in his face that he was enjoying the hike and very much wanted to come along with us, but we parted with precise agreement about where we would meet at nine o'clock the following evening. Then

Doug and I set off down the logging road. Soon Bell Creek met Lawrence Creek, and it, too, gained life as a river, much more impressive than its mundane identity as one of the two principal drainages in the sixty thousand acres of Headwaters.

Just after we left Chuck, we found baby bear tracks hardened into the logging road. I had never seen them before. A baby bear came to life in my imagination and ambled down the road with us. Then mother bear tracks appeared and intermingled with the baby's, until both reentered the pampas grass that lined the road.

Pampas grass. In the setting sun the entire slope above us glowed with such magnificence that one wondered if MAXXAM was growing it as a cash crop.

"You should take a picture of that," I advised Doug. "That's incredible."

"I've got all the pictures of pampas grass I need."

We joined the main haul road that runs along Lawrence Creek. Like its counterpart along Yager Creek, the road was built for speed. At least it wasn't muddy, but its hardness made my joints ache. As we hiked, a huge full moon rose in the east.

Douglas firs

By chance, each of my trips with Doug have been during the full moon. I watched this one rise through an utterly clear sky, inhaled deeply, and gave thanks for the beautiful weather.

"If only Charles Hurwitz were out here watching this moon," I thought, as we walked parallel with Lawrence Creek. But as we hiked, I became aware that the ridge tops were sparsely vegetated. The moon defined the grim silhouettes of only solitary trees. I turned slowly to count how many trees I could see in a 360 degree radius. Twelve. They no longer had elegance. One wondered why they had been spared at all. Seed trees? They were highly vulnerable to being blown down. They looked like mockeries of the original forest that stood just a decade ago.

We turned on the road to Shaw Creek Grove and began to climb. We stopped to rest and sat down in the middle of the road because the last trucks had left for the evening. Doug pointed out a thick stretch of residual old-growth.

"That's an old-style PL thin. See, they thinned out the biggest trees and the second growth is filling in under the old growth they left."

In the silhouette on the ridge, the limbs of the tallest trees still touched one another. Granted, I was looking at a scene several trees deep, but the effect was one of man and nature living in harmony.

"I *like* the looks of a well-done thin," I said. "It makes me feel hopeful."

At that moment, a spotted owl hooted from the mountains to the west of Lawrence Creek.

Doug hooted back.

The owl hooted again.

Doug hooted. At this point in our history, when the forest has become so remote from so many people's lives, it was hard to tell who was more mysterious, the owl or Doug.

The fog moved over the ridge and suddenly obscured all the trees on the ridge.

"Now you see 'em, now you don't," I said.

"Yeah, that's the way it goes around here."

CHAPTER 54 SOUL MATES

WE got up and walked the rest of the way to Shaw Creek Grove. It is hard for me to say how far we walked, because it was so hypnotic watching the solitary trees in the clearcuts come and go through the thick fog while Doug began to spin out one of his engaging stories.

Doug always speaks in a very upright tone that makes whatever he is talking about seem momentous and profound. He could tell you about changing the oil in his van, and you would feel as though you were witnessing history in the making. This time he began telling me about a lawsuit he had filed to try to save the forest that once stood along a stretch of Yager Creek. "Someone told me that PL had filed a plan to log the north side of Yager that included a stand of ancient trees where I personally had seen eagles roosting, watching for salmon in the river below. I couldn't believe that they could get the plan approved, because it would take those trees and all the others that stood close to the creek. I hiked out there and found that they had actually already begun cutting. This was the first anyone had ever seen of salvage logging on PL land."

"*You* discovered the start of the salvage logging?"

"Yeah, at least it was the first anyone had seen of it. They were cutting the giant trees that the eagles used, and they could very well have been used by murrelets as well."

"Were the trees dying?" I asked.

"Not at all."

"So what qualified them to be salvage-logged?"

"They said that they were shading the haul road, and it wasn't drying out adequately. At that time they were just starting to log in the winter."

"What does that have to do with salvage logging?" I asked, confused.

"It's based on their judgment. I came along and found eighty trees that had been recently cut. Some of the stumps were eight feet in diameter. I went to CDF and told them. I said, 'Those trees are not only used by eagles, they also provide critical shade for the river.' They said, 'Oh, Doug, don't worry about it. Those gravel bars are so big down there, those trees don't shade the creek.' So I went back down there, and I took pictures of the remaining trees casting shade and took them back to CDF to show them."

"Had the other side already been logged?" I asked.

"Yeah. When Pat Higgins, who's a salmon expert, did his reports, he found that 98 percent of the shade had already been removed from the main stem of Yager before the plan was filed. This was some of the last shade. The other side had already been stripped. Just a few decades earlier, the Yager basin had been considered a candidate for Redwood National Park because it was one of the best salmon and steelhead spawning basins left in the world. Historically, before the arrival of MAXXAM, the temperature of the creek would have measured in the 60s. Salmon are stressed above 68 degrees. I measured the temperature at 74 degrees at the bottom of a deep pool when they were first starting to log."

As I listened to Doug's story, I was distracted by its epic proportions, not only of the temperatures and the trees, but of the force of Doug's energy and perseverance. I remembered our hike along Yager to Allen Creek. After we had crossed through Yager Camp and hiked north along the barren haul road, Doug stopped to point out the stumps of the eagles' trees. While I never want to see Yager Camp again as long as I live, this young man, at the age of twenty-four, had routinely trespassed back and forth through it, first discovering the logging of the riparian trees; then going to CDF to alert them of this violation of the Forest Practices Act; going back to obtain photographic proof that the trees shaded the creek; measuring river temperatures; and all of this alongside one of the two main haul roads of an exceptionally defensive timber company. I felt as if I were watching an Olympic trial on television. Only this was a different kind of Olympics. It was a trial of responsible citizenry.

Doug went on, "I spoke to a Fish and Game biologist about the danger of the impending plan to the salmon. He said that they couldn't stick their necks out for salmon because salmon hadn't been listed as threatened or endangered yet. So I said, 'What about bald eagles? They're listed as threatened. Without salmon there won't be any eagles.' He said, 'That's a good point, Doug, but we made that connection before about the salmon and the bald eagle, and we got burned on it, so we're not willing to stick our necks out again on that either.'

"So," Doug went on in his understated way, "I began a lawsuit, *Thron v. Pacific Lumber*, to stop the plan."

Of course, I thought to myself. What else is a twenty-four-year-old going to do but raise seven thousand dollars from slide shows and a fund-raising letter, hire a lawyer, and commence suing a multinational corporation? The suit took six months. EPIC stepped in and helped, contributing over ten thousand dollars.

"What happened?" I asked.

"We lost. A guy from Water Quality had done a nonconcurrence on the plan, but Governor Wilson's office had come down on them and pulled the nonconcurrence. It's the governor who ultimately controls most of these agencies. Never mind that MAXXAM was one of the top ten contributors to Wilson's campaign for governor, and Wilson kicked off his campaign with a dinner in Scotia."

"Because none of the local judges like to hear these cases that concern Pacific Lumber," he went on, "a visiting judge from Siskiyou County heard the case. He said that since the agencies didn't have a problem with the plan, he didn't either. After the trial, people saw the judge go up to PL's attorney and pat him on the back and ask when the attorney could come out to his ranch for a barbecue." As clichéd as this may sound, this instance was later repeated to me by one of the witnesses.

As we walked on, Doug began telling me another story that unexpectedly looped around and joined the first.

Doug began telling me about meeting a third-generation PL employee and second-generation old-growth faller, widely recognized as PL's best. Yet he'd been fired. MAXXAM had instituted changes, such as ordering that huge trees be cut closer to the ground to obtain a slightly longer log. That necessitates that a faller kneel with his gigantic saw, the teetering tree blocked from view by the rim of his hard hat, precisely when he needs to be "reading" a tree and getting ready to run This logger began speaking out and was fired.

He has brought suit against MAXXAM, charging unsafe working conditions and wrongful termination.

"I liked the guy a lot," Doug said. "He was finally someone I could talk to, who knows the land the way I do."

His eyes were bright as he described this unexpected encounter with a soul mate. "He enjoyed it too. We even knew the same trees. He was the faller who cut down the trees on Yager Creek . . ."

"What?" I blurted out.

"Yeah, he was the one who cut down the trees along Yager," Doug repeated, his eyes still bright.

What strange ways the human heart works, I thought. I understood how Doug's desire for camaraderie was satisfied by exchanging stories with this man who, though he had not actually accompanied Doug on his trips, had walked the same terrain, days before, days after, years before, years after. But I was appalled that Doug could compartmentalize his passion for Yager and the eagles in his enthusiasm for meeting a fellow woodsman. He and the logger were like explorers, chancing to encounter one another in a wilderness. If one represented Spain and the other England, this did not matter. They were fellow explorers. In their souls they belonged to the same tribe.

But never having met this man, my loyalty was still with the trees, eagles, and salmon. I had seen the line of stumps. I had seen *every single tree* gone. I had seen Yager Creek from the air, flowing naked through a barren, treeless wasteland. And I have not yet looked into this man's eyes and seen his humanity. I can rationalize that technology allows horrendous deeds to be carried out without time for feeling. But I was still chilled by Doug's account.

"He told us that there was a place on Yager where the salmon used to be so thick that the loggers would throw pennies to them and they'd come up and eat them.

"He said that he once cut down an old hollow redwood, fifteen feet across, that was filled with bees and honey a hundred feet off the ground, armloads of it. They put that section of the tree on the truck as a joke, so that the men at the mill would get messed up in it. But the comb blew out on the way there."

I listened dumbly. Environmentalist and timber worker had obviously been like boys together, exploring the woods in their minds, carefree. Though I knew that Doug told me these stories with an awareness of their concurrent thread of tragedy, that the eagles had lost their roost, that there were such quantities of penny-swallowing salmon, that a magnificent hive of bees was destroyed, and though I knew that the logger probably had his own feelings of loss, I marveled at their access to innocence. It made me feel old.

I let my feelings pass and complimented Doug on his lawsuit by emphasizing that I admired his sense of personal responsibility.

"It occurs to me over and over that this is a book about what it means to be an American citizen," I said.

"Do you know who gets madder than anyone at my slide shows?"

"Who?" I asked.

"The veterans. Some old-timer will stand up and say, 'I fought for this country. What are they doing allowing this guy to break the law?' "

CHAPTER 55 SHAW CREEK GROVE

TOM Herman said to Doug during mediation in Doug's lawsuit, "I hate old-growth trees. They are just a pain. I wish they were all gone."

Last night I entered Shaw Creek Grove, exhausted from resisting the sight of uninterrupted clearcuts. We arrived at midnight. Doug had told me this is his favorite grove. Instantly it was mine. The logging road simply ends, butted up against primordial wilderness.

Doug immediately climbed up into the forest and then he called out, "Joan, want to see the salamander pool?"

I was skeptical. In Allen Creek Grove, David and I had bushwhacked for three hours looking for Doug's waterfall.

"Do you actually see it?" I asked.

"Well, not exactly."

"Call me when you see it," I said, and I began to lay out my ground cloth and sleeping bag.

"Come on, I'm almost there."

Curious, I climbed up to the sound of his voice, even though I was envisioning us in three hours, deeply lost in the dark forest. Then Doug's flashlight suddenly caught the gleam of water.

"Here it is!" he said, sounding surprised. "I found it!"

On the hillside an ancient redwood had fallen, perhaps hundreds of years ago, and in the cavity left by its roots, water had filled from a spring. The pool was about four feet across by eight feet long.

"And there're salamanders too!" Doug exclaimed, surprised and pleased. He has told me about the salamander pool for a year now, but more often than not in nature, a place we remember from the year before has changed. As I looked down into the pool, however, I wondered if it ever changed. I began to wonder just how long ago the tree fell. A hundred years ago, two hundred years ago? The soil underside of its roots was thick with shadowy moss. Doug's light panned across the pool and revealed a primordial scene, yellow-gray salamanders casting gray shadows on the soft yellow-gray floor of the pool. Two to three inches long, the salamanders hung motionless, oblivious to being counted for, probably, the first time in their lives. Were they sleeping? Their eyes were too small and squinted to tell if they were open or shut. They were impersonal salamanders, more ooze than flesh. I wondered how many generations had matured in this pool.

". . . eighteen, nineteen!" Doug whispered, hushed by his own awe. "Nineteen baby Pacific giant salamanders! Last year I only found fourteen!

"No, look, there's twenty . . . twenty-one!" he corrected himself. Then he fell silent. "Twenty-two! Twenty-two baby Pacific giant salamanders!" He was relieved. He hadn't talked much about it, but I knew he had been afraid the pool had been destroyed by salvage logging. He was afraid he would climb the slope and find that the very tree that formed the pool had been ripped free and dragged off on a truck. But salvage logging has not yet begun in Shaw Creek. That it ever could seems unimaginable, despite the fact that clearcuts spread on all sides for miles. Intellectually one knows that the surrounding desertlike landscape was once thick with trees, but looking into the complex, interwoven life of the forest, disturbance seems impossible.

"Twenty-three!" Doug exclaimed, still searching the pond with his flashlight. "Twenty-three baby Pacific giant salamanders!"

Pacific giant salamander

CHAPTER 56 SALAMANDERS

TO tell the truth, before I went, I did not want to go to Shaw Creek Grove. Doug kept saying, "You have to see all the groves." But I kept putting him off. So many of our hikes have been so long and grueling and rain-soaked that I heard Ronald Reagan's words, "You've seen one redwood, you've seen 'em all," dare to whisper themselves in my mind. At the same time, I had a feeling that the forest still had something to say, that something would happen that I couldn't anticipate.

By the end of the trip, when Chuck picked us up at nine-thirty at night on Kneeland Ridge, I knew that Shaw Creek was my favorite grove of all, and that our trip had been glorious, with moonlight by night, and dry, pure fall air by day. But I did not feel that something had "happened." I did not anticipate that the experience of the grove would continue to trickle into my soul slowly, the way water seeps into an ancient forest, to be slowly released in springs and seeps. I did not know that the north-facing slope along which we camped would keep appearing to me, on the one hand flat, like the organic shapes of a tapestry, and on the other, like a creature with a soul. I did not know that this week, back in my office, I would become ensnared by the forest's mystery, brought to a new level of understanding and commitment and irrational awe.

Two Pacific giant salamanders, who know no such words for themselves, who do not know in human terms that they are giant, though each is nearly a foot in length, find one another on a rain-slicked fallen branch in the forest. The male with his body says "Follow" to the female, who knows in her flesh to follow. Can she know that to follow means to receive the sperm she needs to fertilize the eggs that she will guard, alone, fiercely, beneath the surface of the pond for nine months? Who knows? Who would think that a female salamander would guard eggs for nine months, the same gestation period as a human fetus . . . but this gestation will occur outside of her body, her eggs hanging like quarter-inch ornaments from the underside of the root bole that roofs her underwater hideout.

The female follows the male, close behind, and when he deposits a neat packet of sperm from his cloaca onto the log's surface ahead of the female, she moves forward, feeling the packet slide down the underside of her own body, until her own cloaca fits over and lifts the capsule of sperm within her. She retreats to lay her eggs, and over the next nine months, if the male attempts to enter her underwater cavity where their eggs hang suspended, or if another male or perhaps an aquatic garter snake should attempt to slide past the female to eat the nourishing eggs, the female salamander will raise up on her toes with her body arched high off the floor of the pond, lashing her tail and growling. If the intruder does not retreat, but advances to test her defense, she will lunge forward, mouth open, and sink her teeth into its skin and tear away a chunk of flesh.

When the larvae emerge, after nine months of care, they will each be only half an inch long. The female will guard them for a month longer, ten months in all, almost a year, before the larvae move away from the nest and forage among rocks and roots, hidden from snakes and water shrews, mink and raccoons that forage, watching, feeling for food. Meanwhile the female, released from her instinct to protect, is free to roam the forest floor again, devouring the slugs and snails, the other amphibians and snakes, even mice and shrews, that chance to be in her path.

I am finally learning the beauty of that which is north facing. Lizard that I am, I have always sought places with southern exposure, boulders and porches, but I am learning that north has its own poetry. I have been engulfed by my growing love for lichens, moss, the tangle of understory that is preserved by the consistent cool temperatures that exist year round on a shaded north slope, a millennium of uninterrupted growth. We can wander all night and all day among the damp cells and molecules of the forest; with our microscopes we can explore the bodies of its creatures, but we will never know the depth of the complexity with which we are privileged to coexist and which we are privileged to destroy.

In Shaw Creek I finally took to heart the waste and inadequacy of even considering the Deal. How many dollars have been paid out for countless politicians and government employees to consider and debate an insufficient solution to the "problem" of Headwaters. The Deal would save only two groves, Elk Head Springs and the grove called Headwaters, leaving the other four, including Shaw Creek, unprotected. It would not provide for the rehabilitation of the devastated watersheds that drain into Yager Creek and Lawrence Creek. The idea that we, as a nation, might concede to pay $380 million to a person whose actions, by most accounts, appear to be unlawful, feels like paying off a terrorist who holds the salamanders, the flying squirrels, the trees themselves for ransom. I felt that this precedent might ultimately face more creatures with extermination nationwide than it might save. We must open our minds to the fact that no more ancient trees should be cut, regardless of where they are.

CHAPTER 57 RELICS

IT is through the bodies of salamanders that the energy of the invertebrate world, that is, the plankton, insects, worms, slugs, and snails that the salamanders eat, is made available to the animals farther up the food chain that eat the salamanders, such as snakes, fish, and birds. Because Headwaters is off-limits to most people and relatively unexplored, no herpetologist I spoke with felt prepared to say conclusively which amphibians inhabit the six groves of Headwaters. The best guess seems to be that in addition to the Pacific giant salamanders, there are most likely southern torrent salamanders, rough-skinned newts, ensatinas, California slender salamanders, tailed frogs, western toads, Pacific tree frogs, red-legged frogs, foothill yellow-legged frogs, clouded salamanders, black salamanders, and perhaps Del Norte salamanders.

Many of the smaller salamanders are uniquely designed to thread themselves through the tunnels of boring insects that riddle rotting logs, devouring the creators of those tunnels and the other animals that make use of that hidden habitat. Tadpoles and other amphibious larvae vacuum freshwater pools and creeks of their algae and other aquatic plants, not only controlling growth but converting energy for use by tadpole-eating animals, including fish. I think of Doug's eagles and how he reminded the man from CDF that there wouldn't be eagles without fish, and I wonder, will there be fish without salamanders? Will there be salamanders without logs?

In the past, many salamander populations were quite large. In one count of ensatinas made in a redwood forest near Berkeley, California, in the early 1950s, there were estimated to be six hundred to seven hundred ensatinas per acre. In many forests there may be more amphibians, by count and volume, than mammals.

The tailed frog and the torrent salamander are considered to be relic species, because they are dependent on conditions that are found almost exclusively in the ancient redwood/Douglas fir forest.

David and I have found a torrent salamander, and as I hold it on my damp hand and look into its beautiful eyes, I realize that the issue of Headwaters belongs to this salamander, whose relatives have lived in Headwaters for millions of years, even more than it belongs to my children, and their children.

In relation to the individual trees around us, which have stood, some of them, for two thousand years, the adult humans of our decade are but a busy whirl, one life stage of one species, endangered perhaps, trying to keep a grip on the big picture. In relation to an ecosystem that has functioned seamlessly for millions of years, these individual trees are but witnesses to quicksilver decisions. Ironically, the trees themselves do not have choice. They stand. They cannot leave, they cannot vote, they cannot protest, they cannot speak words of wisdom, at least not in normal English. Meanwhile, adult humans sculpt the future.

Has Charles Hurwitz ever looked into the eyes of a torrent salamander? I draw the tiny gleam that shines just below the salamander's upper eyelid. The eye itself is black, deep, like the clear spring we borrowed him from. I could show anyone how to create this gleam of life in the eye. The gleams of light that reflect from the damp body of the salamander are made the same way. One simply outlines the gleam with a pale line and fills in the color of the surrounding area, leaving the white of the paper for light. The stars in a dark sky can be created the same way, simply by leaving the stars alone and coloring in the surrounding blue-black of infinity.

I could show how to mix the subtle shade of orange of the adult torrent salamander's underbelly. It would be more yellow than red, with a dulling of blue so that the belly is not too bright, suspended in the waters of a spring or nestled in the moisture of a seep. This is not a flashy animal, but one that relies on the subdued and stable world of the headwaters of streams, where there are no great confluences, but just the subtle oozing of the spongelike land. Beneath the layers of vegetation of the forest, protected by the canopy and then the understory and then the ferns and tangles of rotting limbs and decaying logs, the torrent salamander forages only short distances from the spring, elbows out, stalking the tiniest insect prey, his pale orange belly barely lifted above the moss as he walks.

Rhyacotriton, *this is his scientific name.* Rhyaco *means "small stream," and* triton, *god of the sea, because the salamander is the god of its own tiny sea, the first springs and rivulets receiving water from the upper reaches of a watershed.* Rhyacotriton, *however, is bound to that over which it is lord. The very stability of the environment makes the salamander intolerant of change, dependent on very precise though enduring conditions. Put the salamander in a bucket of water from its own spring and raise the temperature a few degrees and it will die. Cool water, filtered through soil, seeps purely over clean gravel. Here, within the silt-free crevices of the gravel, the female salamanders have laid their large, white eggs layered in clear jelly, in a world with little variation year-round, decade-round, eon-round for millions of years.*

David points out that the torrent salamander's feet are easy to draw because it does not have an opposing thumb. No need to grasp, I think to myself. I look at its feet, and they are like the nubs of fingers on the hands of a human fetus. The whole animal is very fetuslike, though camouflaged to the woods. David draws the tail, and we return the salamander to its pool, where it disappears back into the unknown.

Two hundred years ago, when the redwood forest spread unbroken for a thousand miles, there were countless cool, north-facing slopes and clear, silt-free creeks beneath dark understory that parted only for an occasional upland prairie. But now these salamanders are confined to ever-shrinking islands of forest. They do not know it. The few individuals that survive live as their kind has lived for centuries, but those people who sneak through the darkness with them, carefully lifting rocks, parting leaves, passionate about that which is perpetually damp and cold-blooded, know that all around, the forest, the seeps, the clear-running creeks, are being laid to waste. As we walk on hard-packed logging roads, Doug and I come to pools on the roadbeds fed by ancient springs. We look at each other and think the same thoughts. Each day, tires carry the weight of giant logs through the pools, flattening their sides, compressing their floors, but the water keeps emerging from the ground, because this is its nature. Unlike the amphibians, it cannot die.

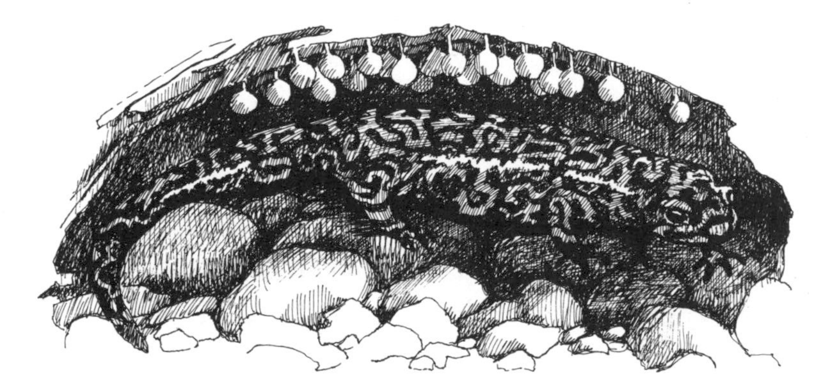

Pacific giant salamander with eggs

CHAPTER 58 EXPERTISE

I RECALL the conversation I had a year ago with a biologist regularly employed by MAXXAM. He is not a herpetologist. He does owl surveys. Perhaps it is unfair to expect him to know about amphibians. But he initiated our conversation because he knew that I was just getting started on this book. I had not yet hiked in any of the groves, though I had flown over the land in a plane. My firsthand knowledge was based primarily upon what I knew of the redwoods in the parks, mixed with simple common sense.

"You know, Joan, clearcuts are actually good for the forest. Sunlight makes the trees grow."

"Well, in limited amounts," I qualified. I was still shocked by the sweeping clearcuts I had seen from the air. "Sure looks like it's hell on salamanders," I supposed aloud.

"Well, they just go somewhere else," he said matter-of-factly. He was someone with whom I had worked on various environmental projects for years.

"Somewhere else?" I asked.

"Well, yes. Like the creeks."

I imagined a salamander trekking across a two-mile clearcut toward a new drainage, dust at first clinging to its damp sides, and then its skin drying out as it scaled yet another dirt clod and descended, crossed the clod's momentary bit of shade, only to scale another, the trip made twice as long by the roughness of the terrain. In my mind I saw the denuded banks of Yager Creek, that, from the air, looked more like a Southern California wash than a river through a rain forest.

"From the air it looked like the rivers and creeks are pretty well stripped."

"Oh no, there's shade," he reassured me with expertise.

Did he actually believe this? He seemed to. Or was he simply saying the emperor wore clothes in order to keep his job? I did not yet have the knowledge or the confidence to survive an extended debate, so I dropped the issue, but my world realigned itself just a little.

Captions follow the color section.

3

4

5

6

7

10

11

13

14

15

16

18
19

23

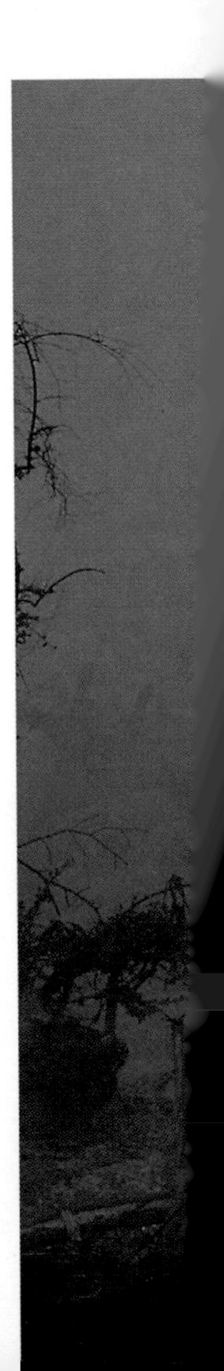

24

25

26

27
28

29
30

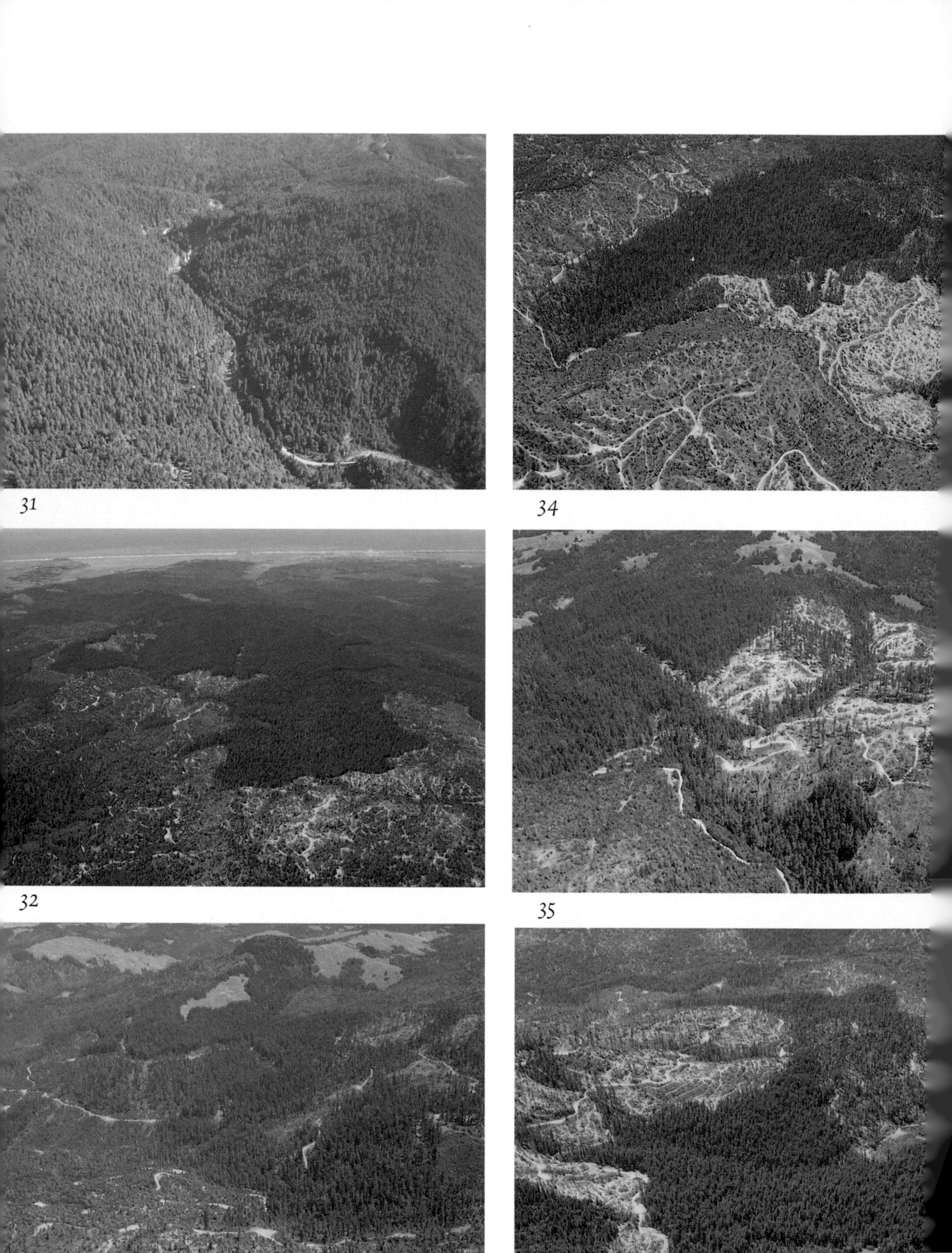

31

32

33

34

35

36

37

38

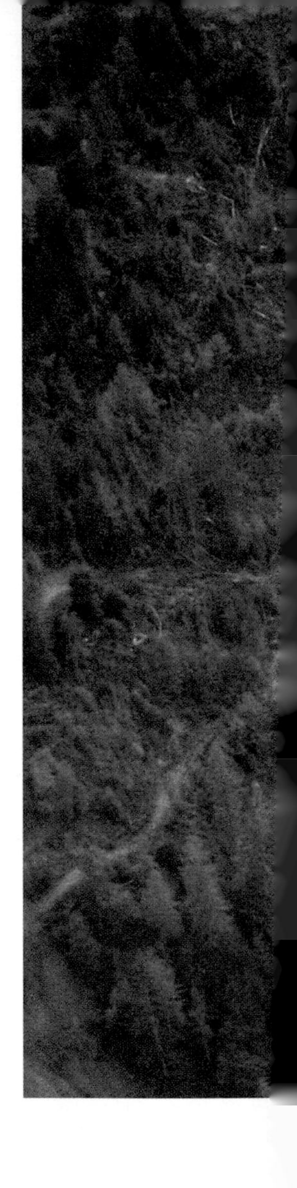

39

40

41

42

43

44

STOP
SACRIFICING
VIRGINS!
OZARK
ENVIRONMENTAL
AWARENESS
PROJECT
60,000 ACRES

45

46

48

49

52

54

53

5

Captions for Doug Thron's Photographic Journey Through the Headwaters Forest

1. *The Headwaters Forest is full of natural wonders like this pristine waterfall in Allen Creek Grove.*

2. *Huge redwoods such as this giant are known to Joan, Doug, Julia, and others as "grandmother trees."*

3. *In Owl Creek Grove, it is common to find the spotted owl, a protected species native to Headwaters and the source of considerable controversy.*

4. *Headwaters Grove. The fog plays a key role in the redwood forest, as moisture is "raked" from the air by the billions of redwood needles and then falls to the ground, in this way accounting for 30 to 40 percent of the annual precipitation.*

5. *Headwaters Grove. The North American coastal redwood forest is the densest forest on the planet, with up to nine times the plant mass of the tropical rain forest.*

6. *Mycorrhizal fungi provide crucial nutrients to the trees in the forest.*

7. *Shelf fungi usually indicate the presence of disease, which often kills the host tree, creating snags and downed logs that provide habitat for small animals.*

8. *The scale of the redwood forest is awe-inspiring—even the ferns dwarf a human figure.*

9. *The cathedral-like beauty of Shaw Creek Grove. This is a north-facing forest that stays damp and cool year-round, making it a haven for amphibians, including the Pacific Giant salamander.*

10. *Boot Jack Prairie in Owl Creek Grove. This is one of the last upland prairies left in the world, completely surrounded by ancient redwood and Douglas fir forest. Doug has seen bobcats, black bears, redtailed hawks, pileated woodpeckers, deer, reptiles, and amphibians here.*

11. *Spectacular sunsets are typical of summer and fall in Boot Jack Prairie in Owl Creek Grove.*

12. *Yager Creek, like many of the "creeks" in the Headwaters forest, deserves to be called a river.*

13. *Because of its steepness, Allen Creek Grove has many waterfalls.*

14. *Near the source of Little South Fork of the Elk River, in Headwaters Grove.*

15. *The area drained by Strongs Creek has mature second-growth forest that is quickly being clear-cut by MAXXAM.*

16. *Headwaters Grove. Soon after a tree falls, it is rapidly covered by various understory plants.*

17. *All Species Grove. Known as the tallest trees on Earth, ancient redwoods may reach heights of more than 350 feet.*

18. *Doug Thron looking at the stump of one of the ancient trees along Yager Creek, where bald eagles used to roost. Doug sued PL* (Thron v. Pacific Lumber)*, in an attempt to protect some of the last old growth redwoods in this part of the Headwaters Forest.*

19. *Ancient redwoods stacked in All Species Grove, after clear-cutting.*

20. *An 8-foot-diameter redwood in All Species Grove that Doug witnessed being cut down. At the turn of the century, it could have taken two men up to three days to cut down a similarly sized tree, but it took one man 45 minutes to fell this giant.*

21. *Doug kneeling down in front of ancient redwoods and Douglas firs cut in Allen Creek Grove.*

22. *A clearcut on the edge of Headwaters Grove, logged by Elk River Timber Company. This is where Ellen Fred did her tree-sit.*

23. *Severe erosion causes road failure, which leads to sedimentation downstream and also allows non-native vegetation such as pampas grass to thrive.*

24. *This erosion resulted from clear-cutting on Yager Creek. MAXXAM's practices have resulted in 98 percent of the shade canopy being removed on the main branch of the river, often making summer water temperatures lethal in this historically prime salmon-spawning basin.*

25. *Herbicide spraying upslope of Shaw Creek. In order to get the herbicides to adhere to the plants, MAXXAM uses thousands of gallons of diesel fuel as a "vehicle." Those using diesel in this way are not required to reveal how much fuel is dispersed.*

26. *Slash burning typically done after clear-cutting on timber company lands. It was a sunny, blue-sky day, and Doug had a brush with death taking this photograph.*

27. *The grove to the left in the photograph is Headwaters Grove; to the right is Elk Head Springs. In the foreground are four square miles of MAXXAM clearcuts.*

28. *Salvage logging by MAXXAM in the Elk River drainage.*

29. *Until 1990, this was a beautiful ancient redwood grove called Murrelet Grove. The creek to the left in the photo was shaded by foliage, and flowed with the clear, cold water essential for the survival of salmon. Now that the trees have been cut, the creek is a huge, exposed canal carrying mud and silt down to the Elk River.*

30. *The devastated Lawrence Creek drainage. More than fourteen square miles of liquidation logging can be seen in the photograph. When MAXXAM took over Pacific Lumber, this area was completely blanketed with ancient and residual forest.*

31. *To the left in the photograph is the 480-acre Allen Creek Grove, the last intact valley in the Headwaters Forest. The area to the right was harvested by Pacific Lumber, prior to the MAXXAM takeover, in an old-style selective cut.*

32. *The 3,000-acre Headwaters Grove, with Humboldt Bay and the Pacific Ocean in the background.*

33. *The 450-acre Owl Creek Grove, with Boot Jack Prairie in the upper center of the photo; the stripped areas have been intensively sprayed with herbicides.*

34. *The 480-acre Elk Head Springs Grove, surrounded by MAXXAM clearcuts. This is a typical example of the severe soil disruption caused by bulldozers during the construction of fall beds, the skidding of massive logs over the ground, and the practice of running bulldozers blade-down, in order to brake. Wildlife in the forested area isolated by clearcuts are exposed to increased predation, wind, and temperature extremes.*

35. *The 400-acre All Species Grove contains rare upland ancient redwood forest. This grove has been severely logged by MAXXAM, first selectively cut, then clear-cut, and recently salvage logged.*

36. *Shaw Creek Grove, home of the pool where Doug and Joan found Pacific Giant salamander larvae. An average of only 60 percent of the canopy still covers Shaw Creek, scarcely enough to sustain coho salmon. Without public acquisition of the 60,000 acres of remaining Headwaters Forest, this population and its habitat will be lost.*

37. *MAXXAM's Pacific Lumber mill in Scotia, California, where visitors can take the company's public-relations tour.*

38. *Logs awaiting processing at the Scotia mill.*

39. *Helicopter logging in the Freshwater drainage. MAXXAM used helicopters to log above the town of Stafford after the disastrous mudslide.*

40. *One of the buried Stafford homes, surrounded by logs and mud.*

41. *The clearcut that caused the mudslide that destroyed six homes in Stafford. MAXXAM logging practices and road-building frequently result in slides such as this, though generally it is wildlife, not people, who are the victims.*

42. *A child's tricycle buried in debris by the Stafford mudslide.*

43. *Hundreds of police confront the thousands of peaceful demonstrators gathered to hear music and speeches at the Headwaters Rally on September 15, 1997.*

44. *Protesters rallying against the Clinton administration's Headwaters "Deal," in downtown Arcata, California, in October, 1996.*

45. *Drummer Mickey Hart and other musicians, performing at the 1997 Headwaters Rally.*

46. *Musician-activists Bonnie Raitt, Mickey Hart, and other demonstrators, at the 1997 Headwaters Rally. These annual rallies have become a gathering place for forest activists from throughout the world.*

47. *Joan climbing the 200-foot redwood known as "Luna," where Julia "Butterfly" Hill set up an encampment in late autumn, 1997.*

48. *Julia Butterfly sitting among the redwood branches near her platform, 180 feet up, overlooking the Eel River Valley.*

49. *Julia Butterfly and Joan, during Joan's and Doug's visit to the Luna platform in April, 1998.*

50. *All Species Grove. One of the most important characteristics of an ancient forest is the multi-layered canopy, which provides diverse habitat and protection for wildlife.*

51. *Along with redwoods, giant Douglas fir trees are the other predominant species in the mature climax forest of the Headwaters. The photograph shows the 137-acre Booth's Run Grove.*

52. *Sawyer in Bell Creek, videotaping helicopter logging in the distance. Sawyer is on an 80-foot-long traverse line between two giant redwoods, which* MAXXAM *has been unable to fell because Sawyer and others are occupying the trees.*

53. *Another ancient grove of trees in Bell Creek that has been protected from logging by the presence of tree-sitters. A tree-sitter with a blue tarp can be seen three-quarters of the way up, in the large spike-topped redwood.*

54. *Sawyer on the traverse line, about 150 feet above the ground.*

55. *Vine maples in the Headwaters Grove.*

56. *Bell Creek.*

57. *A sign of hope. Doug led a group of tree-sitters to this tree near Bell Creek in May, 1998. The next day,* MAXXAM *loggers scheduled to cut down the tree abandoned their plans because the tree was occupied.*

CHAPTER 59 CAMOUFLAGE

I PACK my brushes and a small set of watercolors. Funny to be taking the forest's pigment to it. Red, yellow, blue—red and yellow dulled with blue to make the trunks of the redwoods. I mix the color of the foliage with my eyes—yellow and blue dulled with red—the poetry of the primary colors, all of creation in a balancing act of just three colors.

I choose my brushes. I draw the fine sable of one brush through my fingertips and squeeze slightly to pull them to a point. I moisten my fingers with my tongue and pull on the brush again. A point. I choose four brushes in a range of sizes, ending with a brush for washes. Not too many large areas in the woods, but I'll take it. A small water cup, a single paper towel, a small sketchbook, and my pen, which is like my voice, which speaks from my heart without words.

Later, I get out my camouflaged jacket and laugh at the generic greens. I would like to paint my own camos. I would paint lady fern and sword fern and white trilliums with their three leaves. I would paint a small torrent salamander just crawling over my right shoulder, about to disappear into the dark of my pocket. I would paint the sound of the whirring wings of the marbled murrelet. The ghostly call of the screech owl. The barking call of the spotted owl. I might wash the whole composition out with the blue-black of night and stars, and wash it back in with dawn, filtered through a dripping fog.

Mountain beaver

CHAPTER 60 ELK HEAD SPRINGS

WHEN we returned from Shaw Creek Grove, Doug said, "You know, Joan, now you've seen all the groves but one. If you go to Elk Head Springs, you'll be the only other person besides me who has hiked to all six groves."

I was caught off guard. You mean I had done it? I suddenly changed my estimation of myself. I couldn't believe that in spite of myself I had gained legitimate credentials with regard to Headwaters, not just for sitting at my computer day after day, but for a physical accomplishment. I thought of all the hills I had toiled up, all the wet ferns I had bushwhacked through, all the creeks I had crossed, all the mud I had scraped from my boots, all the times I had hoisted my pack and set off down the road, all the miles I had favored my knee on downhills, all the times I had hit the dirt by the side of the road and scrambled into the ceanothus. Doug had always set my going to all six groves as a goal, but I thought once I got writing, I could weasel out of it. I figured that I could show him that one trip would be enough to report on the experience. Now that I was on the verge of achieving genuine status for all the miles I had put in, I said, "Well, let's go to Elk Head Springs." Then it would be over. Then he would stop bugging me. I would have become the author that Doug envisioned a year ago. To top it off, I had to admit he was right. I can't imagine having written this book without knowing the groves firsthand. This has been the heart of this past year for me. But suddenly I felt sad that Doug would no longer have a compelling reason to urge me to go on more hikes. We have done it. And in the process we have become close friends.

So we made plans. David and Chuck both said they wanted to come along. I got my pack out and began checking off my list: flashlight, Goretex, long underwear, three pairs of socks, fleece. David packs for himself, and I double-check his gear.

I was dreading the actual hike. Few people have ever visited Elk Head Springs, because traditionally it is reached by a long, slogging route up Newburg Road, and then miles past Headwaters. I studied the map and asked, "Why we can't we go in from Kneeland? Instead of turning east to go up to Shaw, we can turn west and wind our way up some of these little roads," I said, tracing my finger along black wiggly lines on a photocopied map.

"No, we can't do that," Doug informed me. "A lot of new roads are not on the map, and we'd be hiking in the dark. Besides, it'd be too far."

I much prefer the drier, warmer country below Kneeland, so I stuck to my idea.

"Measure it. Why can't we just go up here, up here, and then right up here, and we'd be there?" I said again.

"That'd be a steep haul out of Lawrence," Doug countered, beginning to actually study the map.

"It's no picnic up from Newburg Gate. This road looks shorter than that."

"True. Maybe you're right . . ." he gradually conceded. "We'll try it."

A friend of Doug's dropped us off up on Kneeland Ridge, and we retraced the route Doug and I had taken south along Lawrence Creek on our way to Shaw Creek Grove. But instead of turning east again toward Shaw, we turned west and began winding our way up through yet another monotonous maze of logging roads and stripped hillsides.

As we walked, we passed a particularly isolated tree, surrounded by sky and only pampas grass at its feet.

"I don't know why they even leave an isolated tree like that," I said. "Without the forest, it'll probably just blow down. It looks shamed, like it has been reduced to board feet while it is still alive."

We kept hiking, but I noticed Doug had dropped behind. I looked back and saw that he had gone up to the tree and bowed his head, laying his hands on its bark. After he caught up with us again, in a casual welcome, I commented quietly, "We lost you."

"I suddenly felt the tree's pain." He accounted for his absence with simplicity, yet words felt inadequate to express the currents of loss and appreciation that often circle and snake and spiral among us, through us, like vaporous humors in that desertlike landscape. We are so small and our language so crude, we can only fumble to express feelings that still lie beyond our grasp.

We stopped to check our whereabouts on the map, when Chuck scanned the forest with his flashlight and suddenly said, "That looks like old-growth! What trees are those?"

"I don't know," Doug admitted. There simply is almost no old-growth in this area that is not contained within the larger groves. We estimated that we still had a few more miles to go to reach Elk Head Springs, but there were so many new roads that were not on the map that it seemed we were in the middle of a confusing maze bulldozed all over the mountainside. It was hard to argue with the presence of ancient trees, however.

"Could this be Elk Head Springs already?" Chuck wondered aloud.

"What else could it be?" I asked.

"Oh my God, this *is* Elk Head Springs!" Doug said, starting to laugh with pleasure.

I felt like an explorer who had found a quick route to the Orient.

"I can't believe I've never gone this way before," Doug said. "All these years I've only known of one other group of people besides PL employees who have seen this grove, because it's always been such a hard trip."

After the difficulty of reaching the other groves, this easy arrival felt like a gift that was somehow fitting.

CHAPTER 61 DELUXE

ELK Head Springs lies in a wide, north-facing bowl, not as steep as Shaw Creek Grove and therefore not as damp and cool, but its spaciousness invites exploration. The trees are the biggest and oldest that I have seen in any of the groves. The understory, on the other hand, is not as thick as in Shaw Creek or Headwaters. In places the sword fern seems almost parklike rather than overwhelming

But for me it does not have Shaw Creek's wild privacy. Its creek, headwaters of the South Fork of the Elk, already had a layer of silt on its boulders from a clearcut upslope, like an altar that needed dusting. Throughout the grove, biologists' graffiti identified groups of trees with foot-tall white spray-painted numerals. It was impossible to imagine how anyone could spray-paint large numbers on the patinaed bark of a 1,500- or 2,000-year-old tree.

"Someone should find out who did that and go paint white numbers on his car, his trees, and house," Chuck said. David looked up at him in surprise.

"Not really," I quickly clarified.

"They could use water-based paint and at least give him a scare," Chuck persisted. I said no more. David knows that Chuck has better things to do.

As we explored, I found the biggest tree that I have seen on all of our hikes. Not only was it healthy and extremely tall, but its trunk did not taper inward as the tree ascended, giving the tree the extremely massive shape of the *Sequoia dendron*, the redwood that grows inland in the Sierras. The tree stopped me as if it spoke my name. The Dyerville Giant remains magnificent even after death, but it is a public tree. This one few people have ever seen. We estimated that it is about nineteen feet in diameter. Its trunk has an even coating of gray-green lichen over the charcoal of ancient fires. I suggested we sleep below it.

We ate a simple dinner by most standards, instant mashed potatoes mixed with red beans, but for Doug and I, who usually do not even bring a stove, contenting ourselves to sit on our respective sleeping bags munching cold food in the dark, it was deluxe. After dinner, in honor of David, who loves games, we all played five-card draw by flashlight. It seemed appropriate for a ten-year-old to play poker among two-thousand-year-old trees. He will need more than reverence as he grows up. He will need humor, and shrewdness, and courage, and the companionship of men with integrity, and luck.

We lay out in our sleeping bags with stars just barely visible through the canopy. A screech owl called, its eeriness appropriate to the wildness of Elk Head Springs. Then there was silence, holy silence, and we slept.

CHAPTER 62 WEEDS

I WAKE. My eyelids part halfway, and the forest rises three hundred feet above me, lifting my focus from the close world of sleep to the haunt of the owl that last night called. . . . Saws? I hear saws. It is Saturday, a day of silence for the forest, which for thousands of years has heard only the sounds of its own creatures, the wind, the rain, lightning, fire, and perhaps the occasional passage of its native people, born of huckleberry and salmon, acorn and deer. Today we will listen to saws? Why?

"It's on the opposite ridge over there," I hear Chuck say nearby.

I am not yet ready to wake up. I close my eyes.

"It's Saturday. Why are they cutting on Saturday?" Doug asks.

"It will stop," I think, unwilling to admit to a day that will very likely be dominated by the sound of saws.

"I dunno. It just keeps going. Usually they stop, and then you hear the tree fall. It's just been that constant 'whirrr, whirrr, whirr . . . ' since dawn."

I do not want to roam this grove with the sound of saws reaching through the trees until the loggers' quitting time at three-thirty. I will sleep. . . .

"Hear that, Joan?"

"Yes," I answer against my will.

"It's weird. It doesn't stop. It just keeps going."

"I know."

All morning the saws continued. It was so unsettling that we decided to risk crossing the valley to see what was happening. We climbed downhill through the grove, weaving through clumps of sword ferns, and over and around gigantic logs that displayed the infinite number of ways a tree can die and decompose. Some were elevated on other fallen logs, providing sheltered spaces underneath, alive on their upper surfaces with a verdant growth of ferns, lichen, huckleberries, and fungi; others were closer to soil, rotting, giving way dangerously beneath one's feet; still others remained upright as towering snags. A powerful fire had burned through the grove, perhaps within the last hundred years, and each tree had a different story to tell of this event alone. We came to the bank of the south fork of Elk River, its ancient bed deeply incised between steep banks, and I climbed down in my usual fashion, gripping several fronds of sword fern at a time and using this as support to lower my weight. We stopped to fill our water bottles, careful not to disturb the silt that coated the rocks.

To a tourist passing through, the silt might seem like a minor detail, but to a tailed frog or a salamander, it is a matter of life and death.

Unseen, in a crevice between two rocks, two tailed frogs are locked in the act of mating. The male did not sing to find his mate. He has no voice, nor has she. Neither could she have heard him if he did sing, for her species, evolved within the constant sound of the rushing creek, has no need for external ears.

The frogs' lungs are greatly reduced as well, decreasing buoyancy. For twenty-four hours the male and female have hung suspended, coupled. Tailed frogs are the only frogs that engage in internal fertilization, the male inserting his "tail" into the female's cloaca. Otherwise, sperm and egg might both be washed downstream. After the male has released his sperm and his grip, the female will retain the sperm within her body for ten months before the eggs are actually fertilized. In July, she will lay fifty large white eggs underwater in a hidden cavity. While the eggs of a western toad laid in warm water on the margin of a still pond downslope will hatch in only five to seven days, in cold water, embryonic development is slow. The eggs of the tailed frog require seven weeks to hatch. If, in that time, the water barely warms, perhaps not even enough to be detectable to human skin, the eggs will likely die. In a creek where change almost never occurs, flexibility has not evolved.

We kept going and found one cause of the siltation, a raw dirt landing that dumped right into the

Tailed frog

creek. As we climbed the barren dirt bank, leaving the forest and suddenly surrounded by a wasteland of pampas grass, the saws grew louder than ever. I was already sorry we had decided to come. But we continued on.

Out of fear of having his photographic equipment permanently confiscated, as has happened to countless innocent bystanders at different rallies, Doug has conditioned me to be ready to head for cover if we hear the sound of a truck coming our way. This is not exactly my style, so once we joined the main road I was jumpy and watchful.

"Wait! Is that a truck?" Everyone stopped and listened.

"No, it's a plane."

We started walking a ways and someone else suddenly asked, "Shhh! What's that?"

"Nothing."

"What will happen to us if we get caught?" David asked realistically.

"They might take us to Eureka and fine us for trespassing."

The idea didn't appeal to either of us. David and I are still rather conventional, law-abiding citizens with the full set of societal inhibitions, in spite of the fact that we have spent a year hiking with Doug past the most shocking violations of political and spiritual laws that govern stewardship of the land. As I hiked I thought about the fact that Charles Hurwitz's own faith, Judaism, has some of the wisest and most insightful mandates regarding stewardship of nature.

Suddenly a flock of quail burst into the air, sounding a great deal like an approaching pickup.

"Run!" I yelled and beat everyone into the cover of some nearby pampas grass. I peered out and David, Chuck, and Doug, after an initial start, stood doubled up with laughter out on the road, waiting for me to pull myself free from the razor-like leaves.

When we were right below the saws, we sneaked into the underbrush and bushwhacked up to take a look. At exactly three twenty-five the sound of the saws ceased, equipment was packed up, and a truck passed by on the road below. We watched until it entered a clearing on a distant ridge. Three men rode in the back.

As we climbed the slope, we found ourselves ensnared in a no-man's-land of cut brush and toppled small trees. It was almost impossible to move. We stood looking around, trying to find words to interpret the chaos around us.

"Look, they've just been cutting brush all day. That's why we didn't hear any trees falling."

"And that's why they never turned off the saws."

"What an awful job. Oh my God, what are they going to do with this mess?"

"Maybe they are trying to clear a path so that the guys with the herbicide sprayers can penetrate this brush."

We stood longer, each of us slowly turning around, trying to sort out what could have been, what should have been, since the area was no longer vegetated with forest. There is a saying, "The best fertilizer is the footsteps of the landowner." Sheer common sense told me that this land had gotten away from its owner. By clear-cutting so much area eight or ten years ago, and leaving the hillside fully exposed to the sun, rampant recolonization by tan oak, alder, and ceanothus had gone unchecked.

Standing in the midst of such overgrowth, observing the intense amount of labor involved in "manual-release," as this kind of hand clearing is called, the casual observer might be easily convinced that it is good forestry practice to employ herbicides and slash burning to eradicate "weed" trees and brush so that new seedlings get a nice, fresh start without competition. I have begun to refer to this notion as "kitchen forestry." It appeals to the side of us that finds satisfaction, after we have washed the dishes and set the kitchen straight, in making a final pass over the kitchen floor with the broom. This works with linoleum. But giving a final "neatening" to a hillside with a quick petroleum-fueled fire and a dose or two of herbicide utterly thwarts the wonderful resiliency of the forest ecosystem.

Throughout much of the West, when newly cleared soil is left exposed, one of the first shrubs to colonize it is ceanothus, the various species of which are known by the common names California lilac, wild lilac, buck brush, and snow bush. Perhaps I have lost perspective, but I find a comforting poetry in the way that ceanothus functions in the reestablishment of conifers after land has been cleared. One of timber companies' primary complaints about ceanothus is that it grows faster than the conifer seedlings that sprout in a clearcut and therefore shades out many of the young trees. In fact, however, more seedlings usually sprout than can actually reach maturity in any one area due to space limitation. Ceanothus bushes act as natural spacers, shading out the weakest individuals and then dying back once the more vigorous trees overtake them in height. Not only does the expansive root system of the plant help to stabilize and aerate the soil, but, in addition, ceanothus has the ability to fix atmospheric nitrogen in the soil, providing a natural fertilizer for the young conifers.

If beneficial vegetation is routinely eliminated through manual release and herbicide spraying, how will the soil to be replenished with basic nutrients such as nitrogen? Alders, another "weed" in the eyes of timber companies, can fix up to three hundred pounds of atmospheric nitrogen per acre every year. The soil beneath a stand of alders that has been allowed to precede a coniferous forest in natural succession, may, at the end of forty years, contain up to *seven thousand pounds* of nitrogen per acre.

I loved alders when I was a little girl, not only because one usually encounters these pale-trunked trees growing close to streams, but because even though they are broad-leafed, alders bear charming miniature cones. After researching alders, I now feel awe when I drive or hike into a stand. They are perhaps my favorite tree, because they are common and, given time, they are equipped to significantly correct man's appalling capacity to destroy the soil.

Tan oak, manzanita, and madrone, the hardwood "weeds" of the forest, can serve as valuable hosts to mycorrhizal fungi while conifers are becoming reestablished. In one test, scientists located sites in southwestern Oregon that had once been forested with virgin Douglas fir, but which, after they had been clear-cut and burned, seemed incapable of supporting newly planted seedlings. The seedlings always withered and died, leaving the land hopelessly barren. This is not an unusual problem for logging companies, MAXXAM included, so soil scientists were hired to look for a solution.

They gathered soil samples from a variety of the barren sites and also from locations where tan oak, manzanita, and madrone grew densely. They took these soil samples into the controlled conditions of a greenhouse and planted Douglas fir seedlings in each type. After five months the seedlings growing in the soil that had been in contact with the roots of the hardwoods were 60 percent taller, more than twice as heavy, and had almost twice as many roots with mycorrhizae on them as the seedlings growing on soils that had not had their mycorrhizal populations sustained by the "weeds."

As we bushwhacked our way back out of the cut area, I recalled the fact that if the original forest had simply been thinned by 60 percent of its volume, and the growth in the younger trees which had been suppressed by the shade of the taller trees thereby released, good saw logs could have come off that same ground in another fifty years. This was the old PL's system. It was simple and it worked. In a 60 percent thin, adequate shade is left to prevent the brush from ever getting started, the roots and soil remain intact, and, while much of the virgin understory would have been disturbed, at least the forest itself would continue to grow, to the benefit of the animals who live there, the company that owns the forest, and the people who use wood.

We headed back to Elk Head Springs Grove, and as we neared the small abandoned road that dead-ends at the grove, I experienced that leap of the heart that one can feel at the entrance to one's own driveway. I was home.

CHAPTER 63 MORNING

THIS is our last day at Elk Head Springs. I don't want to leave. I now understand how Doug has spent weeks at a time out here in these groves. They feel like the safest places on earth. I wish that we had not taken the time to bushwhack up to the sound of the saws yesterday. I wish I had spent more time simply exploring this grove.

Now I hike alone, beckoned uphill by the warmth and dryness of morning that is dawning on the other side of the ridge. As I hike I avert my eyes from the white spray-painted numbers that mark the trees. I hit a road, punched into the grove, and the words to Judy Collins's melancholy song "The Coming of the Roads," which I memorized twenty-five years ago, return to me in fragments of verses.

"See how they've torn all to pieces, our ancient poplar and oak, and the hillsides are stained with the greases, that filled up the heavens with smoke..."

As I near the top of the ridge, old-growth tan oak begins and the day's warmth reaches my skin, bringing with it the scent of the sunny slope to the south. I want to come back here for a week, not just with David, but with Suzanne. I want her to know this privacy and silence and peace. She too must have it as a point of reference in her life, so that she will recognize, by comparison, what is noisy and unpeaceful. If there aren't people who value silence, who will protect it?

I think of my refrigerator at home and how I breathe easy when it stops running. My soul longs for silence such as this. I think of Jean, my daughter's piano teacher, who is bothered by the sound of a refrigerator as much as I am. I wish she could hike here with me. I wish I could give her this silence as a gift.

"Once I thanked God for my treasure, now with rust it corrodes . . ." I look left and right and realize that if the Deal does not go through, everything 150 feet from the road will be salvage-logged. With all my heart I want to bring Charles Hurwitz here, to his own woods, and have him look and listen. . . .

Then Robert Frost's words "Whose woods these are, I think I know. His house is in the village though . . ." come to me. Robert Frost knew the poignancy of receiving the gifts of another man's possession. If only Hurwitz would make this pilgrimage on his own feet, alone, with no vehicles—feel the ache of the soles of his feet at night and then, in the morning, feel his feet restored by sleep.

I stand up to go. I don't want to go. I look east and thank Gaia and see the beams of the sun coming from behind six snags that stand in a group. I hear a woodpecker working on the side of one of them. I pledge myself to those snags. If salvage logging begins, they will all be cut. The tapping—

loud, hollow—sounds like a lone craftsman building a humble shelter. This is who is working . . . a humble craftsman . . . we are all humble craftsmen.

My eyes follow up the snag to its highest point. I feel like I am at the pinnacle of my life. Thank you, Gaia, for giving me the task to write about this silence punctuated by a tan oak acorn falling . . . the hollow tapping of the woodpecker . . . a fly . . . silence . . . my hair against the neck of my vest . . . the fly again . . . a flicker calling . . . silence . . . and silence . . . another acorn . . . silence . . . the fly again . . . silence . . . a twig falling . . . a squirrel scolding distantly, ticking off moments with its voice . . . a winter wren's call notes . . . chip, chip chip, chip . . . saws? no . . . it is Sunday . . . silence . . .

CHAPTER 64 SOUTH FORK OF THE ELK

WE left Elk Head Springs and headed north toward Kneeland. I was walking ahead. When I am faced with a long hill, I like to just keep walking, slowly, steadily. The others were talking; Doug was stopping to take photographs, so I simply moved on. Suddenly I thought I heard a truck. I dove into a clump of pampas grass and instantly a black pickup appeared around the corner, going very slowly and quietly. I pulled the pampas grass over my turquoise pack and stared out through the leaves. The truck whispered past very slowly, two men dressed in camouflage standing in the bed with guns trained over the cab.

As I write this now I can hear Ralph Kraus, who has been PL/MAXXAM's neighbor for forty years, ask me, "Why'd you hide?" He would never hide on a logging road. When he asked me this, I felt a little silly. Why didn't I just stand my ground? I wasn't hurting anything. I had nothing to hide. Many people have told me that the old PL let them walk their roads or ride horses on them or even drive them. But MAXXAM does have a lot to hide. I guess Doug has trained me.

I immediately began to worry about David, but I could do little except wait to see what would happen. Who were those men? What were they looking for? Then I suddenly realized . . . hunters . . . It is Sunday . . . silence . . . I made myself comfortable and looked up at the sky. Then I heard yelling and a truck door slammed. My heart sank. More yelling and . . . silence. I waited, vowing not to separate again. What had happened? Was everyone all right? The silence continued five minutes, ten minutes. A song sparrow landed in my clump of pampas grass, and I vowed to sit still in "blinds" such as this more often, not to hide from people but to watch birds. Suddenly my concern intensified. Where were they? What if something really had happened?

Just then the truck came whispering back up hill, guns still trained over its cab, this time the men looking around actively. They passed. I waited. Five minutes, ten . . . fifteen . . . and then Doug and Chuck and David came quietly up the hill. I hooted softly and came out, and we walked together until we found a place to get off the road. We were all badly scratched. Pampas grass has razorlike leaves, and we'd all been into Himalaya berries as well. The truck had been moving so silently that it caught Doug and David and Chuck on the road. Doug took off with his ninety pounds of photographic equipment, but Chuck and David hid unsuccessfully in the brush. One of the men was yelling at David, who was still visible. Suddenly, on Chuck's cue, both of them dove straight off the ten-foot cliff behind them, and took off running, catching up with Doug far down the slope. By the time we compared wounds, David had a goose egg where he had hit heads with Chuck, who was already developing a black eye. Just then we heard the truck again. We dropped to the ground and watched. They had circled around on a skid road and were coming back up hill again, guns still trained over the cab.

A friend of Doug's was supposed to meet us at nine-thirty P.M., up on Kneeland ridge, but we realized we would never make it, so we decided to head downhill along the South Fork of the Elk to Kristi Wrigley's house. This involved compass work in order to bushwhack over to land owned by Elk River Timber. We retraced our steps and headed west toward the lowering sun, passing Elk Head Springs and coming to the end of the road. Chuck and David leaned over the map and plotted our route. We would be heading right through a huge tract of land offered to Hurwitz as part of the Deal. We estimated that it would be a seventeen-mile hike in all, not a small distance for a ten-year-old boy.

We dropped into some dense second-growth and then suddenly hit an old logging road that made its way through the otherwise nearly impenetrable thicket.

"We hit our road!" David and Chuck congratulated one another. It was a sweet, old-fashioned road, narrow, winding through the woods. I was already exultant that we had decided to leave the hunters behind and see new sights.

We came to a set of brand-new markers on a tree, which were startling in such an out-of-the-way place, as if aliens had landed and staked a claim.

"What's that?" I asked.

"Those are the markers for the Deal!" Chuck informed me. "Look. It's all been laid out already!"

We had rejoined the South Fork of the Elk, which has its headwaters in Elkhead Springs, and began following it on the grassy, picturesque road. The creek was cobbled, with little evidence of silt. The entire scene was comforting after the unrelieved harshness of the cut-over land, although the markers for the Deal ominously portended further change. It was strange to see signs of governmental certainty that the Deal would go through, when so much controversy still surrounds its passage. We reached the end of the grassy little road where it T'd with a well-traveled logging road. Still spooked by the hunters, perhaps, I thought I heard a truck coming and started to dodge off into the brush.

"You don't have to hide here," Chuck said.

"Why not?" I asked.

"This is Elk River Timber. They don't have anything to hide," he replied.

"How come?" I asked.

"Because they're not practicing liquidation logging."

I am so accustomed to MAXXAM that it felt strange to enter land where logging was perhaps under control. I looked around to see if it looked different and immediately noticed the complete absence of pampas grass. It was like stepping back in time. We hiked to the top of the ridge and stopped, stunned by what we saw.

"This looks exactly like the whole Yager Creek valley once did when it had trees," Chuck said.

"I know," said Doug. "It looks like Lawrence Creek drainage did too, before they cut it."

Until that moment, Doug and I had been at odds with Chuck over the Deal. I strongly felt that with so much alleged corruption gone uninvestigated, Hurwitz shouldn't get one more tree or one more penny from the American people. This was simplistic, but my mind had dug in its heels and said, "It's not right."

Doug, who probably knows MAXXAM land as well as anyone alive, has always felt that we must acquire all sixty thousand acres and the South Fork of the Elk, a stand long advocated by the environmental community to protect whole watersheds, not just isolated groves.

Chuck, on the other hand, felt that Headwaters Grove and Elk Head Springs simply must be secured by means of the Deal, even if it left the other groves and the South Fork of the Elk protected only by the beleaguered Endangered Species Act. When he looked down on the valley of the South Fork, however, washed from side to side by a sea of trees similar to that which he had explored for years in the Yager Valley near his former home in Carlotta, he stood silent, obviously confused and in pain. All of his well-defended stance on the Deal was gone in an instant and, after a brief pause during which he was totally at a loss, his mind began to work in a new direction.

"They can't have another valley," he said. "If they get this, it'll look just like Yager and Lawrence in a few years."

"I know," said Doug, also lost in thought.

It is the continuousness of the undisturbed habitat that is so striking. Most of the valley is large second-growth trees.

"This should be purchased for a park, and combined with Elk Head Springs and Headwaters." We each said this in our own way. It was obvious. Our wish list was getting longer. We came to some selective thins.

"This is how a thin is supposed to look," Chuck informed me.

We came to two small clearcuts, their blackened stumps and logs still smoldering from slash burning.

"What are these?" I asked, suddenly feeling double-crossed.

"They've done high-lead yarding here because it's so steep." This meant that they had set up a cable system to lift the cut logs upslope rather than dragging them. "They don't run Cats," Chuck continued. "They do small clearcuts because of the time and effort involved in setting up the yarder."

I wasn't totally convinced, but in general the scene looked so under control relative to MAXXAM that I didn't argue. What an education David was getting! I imagined him sitting in class in a few years at HSU as the teacher began talking about selective harvesting or high-lead yarding or the meaning of sustainable forestry. Of course, in the way that children always peer to the very core of a parent's soul, David has a passion for the desert and cactus. But life just might fool him, the way it has fooled me. For much of my childhood my brother talked forestry. It is no coincidence that I am writing this book, even if my point of view differs from his.

We all felt hungry, so we sat down in the middle of the road and ate dinner. As we alternately chewed silently or talked softly, it got dark and began to rain. We put on our rain jackets and pants and saddled up our packs and kept walking. It is hard to remember how long we hiked down the ridge. I was in that numb state toward the end of a trip when I just walk. Doug and Chuck noticed after perhaps three hours that we had come back onto MAXXAM land. I looked out at the silhouette of the ridges around me and saw the familiar skyline of single, beleaguered trees surrounded by emptiness. The pampas grass had begun again and lined the road. I was glad it was dark.

Suddenly we noticed tiny lights below us. We had been away from the sight of any lights but our own flashlights for three days, so these looked completely foreign to me. We were above the little community of Elk River, which is tucked back in the clefts between the mountains, where river loam and logging have afforded a living for the last hundred years.

Doug stopped and howled like a wolf.

"Don't do that!" I said in a low voice, not wanting anyone in this peaceful setting bothered any more than they already had been by uncontrolled flooding and siltation. David loved it. We followed the road down to the metal gate, climbed over, and were instantly transformed into normal citizens, except that we were carless and walking down a dark country road with backpacks at nine-thirty at night. We passed a man who was in his living room watching television. How curious it was to see another human. I looked at him as if I myself were not human. We continued on, to be greeted by the dogs of Elk River, who had not been told what to do if four friendly people came through wearing perplexing loads on their backs. Some barked; others circled us sheepishly, anxious for an unscheduled petting.

We crossed the bridge over the main stem of the Elk and headed back up the other side past Ralph and Nona Kraus's house. They appeared to not be home, so we walked on toward Kristi's.

"Where does she live?" Doug asked.

"Up here . . ." It was another half mile out to her family apple orchard.

We went to the front door and knocked, and Kristi welcomed us as if she had been expecting four exhausted and dirty people to collapse in her living room. Ralph and Nona were there. A watershed meeting had just ended.

What would the Deal do to these beautiful neighbors, already overwhelmed by the disruption of their community? Nona has been on crutches for the past several months. I haven't asked her why, but as I looked into her tired, gentle face I thought to myself how much she must want peace restored to her valley and her life. Ralph is a tall, soft-spoken, naturally dignified, retired schoolteacher. I thought of the letters Ralph had lent me describing Elk River when he and Nona first moved there in 1958.

> *At that time Elk River was a clear stream with a fairly good summer flow and supported a sizable population of cutthroat trout, steelhead and coho salmon. Fish could be seen in every pool and there were numerous pools with nicely flowing water and riffles between the pools . . . In 1960 our son was born and he and the two girls grew up enjoying fishing and swimming in the river. They swam in a five- to six-foot-deep swimming hole at Wrigley's and in another larger one about half a mile upstream. Both swimming holes were large enough to allow a rope swing and deep enough that the kids could dive. Our younger daughter and our son often fished the river and usually came home with several nice cutthroat and/or steelhead. . . . In*

1990 all of this began to change. . . . We had as many as 120 loads a day coming down our little road all year long, with each truck carrying about 5,000 board feet of logs. One can quickly calculate the footage coming off of the hillsides of the North Fork of Elk River. . . . The deep holes are now filled in. The bottom, once gravel, is now mud, and the gently sloping banks at the bends in the river are now steep mud banks. . . . In walking a couple of miles of the stream I saw no salmonids, only a few sticklebacks. Even the tributaries that once had small fish in them are now totally unfit to support any fish life.

The Krauses and Kristi live below the confluence of the North Fork and the South Fork. What will happen if the Deal now gives MAXXAM the rest of the watershed of the South Fork to log? What kind of "take" does this constitute?

After filling our arms with good things to eat, including apples from her orchards, Kristi and her sister gave us a ride back to Arcata. I had officially seen all six of the old-growth groves, and most of the rest of the clear-cut land that comprised the sixty thousand acres recommended for protection by those who believe we need viable watersheds and not just isolated forest shrines. I came away from the hike with a new concern, however: the South Fork of the Elk.

Salal

CHAPTER 65 AN APPOINTED TASK

I KEEP being aware that I have not seen a tree cut down. Actually I have. We used to selectively log the beeches and maples in the woodlot on our farm in Vermont for firewood. And we used to selectively cut hemlocks, load them on a friend's flatbed, and take them to the mill for lumber that we needed for the farm. What I mean is that I have not seen an old-growth redwood or Douglas fir cut down. I have seen videos that are being used as evidence in a lawsuit against MAXXAM, *of an old-growth faller cutting down the gigantic trees. But I feel obliged to be there in person and hear the saw and smell the resin. The prospect scares me, but I feel a need to bear witness.*

CHAPTER 66 MAIDENS

RECENTLY the California State Senate held a hearing to investigate complaints by people such as Kristi, the Krauses, and Mike O'Neal that CDF is not doing its job. I went to interview an agent from the Department of Fish and Game who gave meaningful, heartfelt testimony at that hearing. His name is Bill Condon, and I wanted to ask him what motivated him to stick his neck out and risk speaking the truth. As we started talking I was relieved to find out that his world is bigger than just Humboldt County. His credentials include a master's degree in international forestry from Yale University and several years in Nepal with the Peace Corps.

"What's depressing," he said, as he described his experience in Nepal, "is that we do the same thing over and over again. In Nepal an old person standing in the middle of a barren, treeless area said to me, 'Yes, this is where tigers used to roam when I was a boy and there was jungle here.'"

"What prompted you go to the senate hearing?" I asked.

"My interactions with the landowners who are affected by the logging. My sense of outrage at how their legitimate concerns are being responded to by CDF. CDF says, 'We're going to assume that the rules will take care of it.' But they're not. In California we have some of the best forestry laws that exist. There's wonderful 'intent language,' but when it gets down to what is actually happening on the ground, sympathies are with the silvicultural values of the landowner." Of course, "the landowner" is often a multinational corporation, not the little guy downstream.

"Do we need more agents?" I asked.

"Yes, this is a necessary step, but not a sufficient step. The main problem is, this system has no objective criteria for deciding when enough is enough. There are no definitive yardsticks for CI—cumulative impact. It happens one THP at a time, but there's no one looking at the overall effect. Disturbance has always had an important role in our ecosystem—fire, flood . . . Debris torrents are an important source of gravel for fish—but not to this degree.

"CDF has expertise in putting out fires, replanting, generally evaluating silvicultural methods, but the ultimate decision-making process should not be left up to them if the objective is to protect the environment. They do not have the training to assess impacts on biological resources."

"I think most people think that our wild plants and animals are well protected by the Fish and Wildlife Service and the Department of Fish and Game," I said. "How much input do you actually have on protection of the wildlife?"

"Though we are charged with protecting the public trust, we are only in an advisory capacity. Our wardens enforce hunting and fishing regulations and some water pollution laws. We can do something if you drop a battery into a stream or you poach a deer, but we can't do anything if you dump silt into a river, as long as it is the result of an approved THP, and virtually all timber harvest plans are approved.

"In the best of years, Fish and Game only reviews 20 percent of plans that are approved. This year, because we are so busy doing HCPs and SYPs, we have only reviewed five plans in all of Del Norte and Humboldt Counties, and the year is coming to an end."

I should interject that an HCP is a Habitat Conservation Plan, which is a "permit that allows a landowner to 'take' an endangered species if this happens incidental to an otherwise legal activity." An SYP is a Sustained Yield Plan, which is a "document that projects corporate timber production into the distant future." These definitions are from the glossary of the *Headwaters Forest Stewardship Plan*.

"The Department of Water Quality is more active than we are," Bill continued. "They review Timber Harvest Plans that occur in 303-D watersheds, that is, those watersheds that are designated to be impaired due to sediment or high temperatures. Seventy percent of all watersheds are now water quality impaired in Del Norte, Humboldt, Mendocino, and Sonoma Counties."

"Seventy percent?" I asked, incredulously. "What should be done?"

"We need significant structural changes. The Fish and Wildlife Service gets involved only if there is a federally protected species that's involved, like the marbled murrelet. The Board of Forestry writes the rules, and CDF enforces them. The Board of Forestry and CDF need to be out of the driver's seat. As it is now, the public trust resources owned by the people of California are being sacrificed for the sake of jobs and timber harvest. It should be the other way around; the goal should be environmental quality. The timber companies are externalizing the cost of poor practices to the public, herbicides, clear-cutting, loss of the fishing industry, the dredging of the harbor . . .

"We need objective standards to evaluate environmental abuse. Right now it is up to the landowner [again, read, multinational corporation] to determine whether the public can even attend preharvest inspections. During the review of harvest plans, CDF often reminds people that 'this is not a public meeting,' and if they want to they simply clear the room."

"What about hiring someone to oversee timber harvests?"

"In Santa Cruz, they do this. They didn't feel that CDF was adequately protecting them, so they voted to have county rules and they hired Dave Hope to look at every single harvest plan."

"Dave Hope?" I said. "I've spoken with him over the phone. I didn't realize this is his role in Santa Cruz."

Dave Hope is the man I interviewed over the telephone who went on the tour of the Yager Creek logging "show" and sent me the letter expressing his outrage over conditions there. How nice it would be to have local control of our resources in Humboldt County informed by someone like Dave Hope. I recalled a panel of CDF foresters who sat before a meeting of the people of Freshwater not long ago, and, when pushed, admitted that, in fact, they guessed they were powerless to stop MAXXAM. I scanned across the panel of men, estimating what each man earned and began to add . . . $40,000, $80,000, $120,000, $180,000, $240,000, $280,000, $320,000 . . . How many hours have unpaid citizens expended doing the work of these men? I thought bitterly of environmentalists' perpetual struggle to keep vehicles running, to pay lawyers, to pay phone bills, to buy film and print photographs, to print satellite maps, to test for herbicide contamination and sedimentation . . . the list goes on and on. I thought of the lawsuits that should have been undertaken, but weren't for lack of funds. Yet these men sat there, slightly dazed by fear and confusion over the threat of losing their jobs, saying, "We just don't have enough science to prove that clear-cutting causes adverse impacts." Meanwhile the trees keep falling.

Bill continued, "Tom Herman, quoting MAXXAM's fisheries consultant, said, 'There's no crisis.' For MAXXAM there is no crisis. But with the cumulative impact of THP after THP, there's a crisis for people like Kristi."

Cumulative impact, cumulative effects . . . these often "run downhill." There's not only a crisis for Kristi, there's a crisis for salmon, and salmon have played so massive a role in our ecosystems that I predict they will be missed by more than just the salmon fishermen. Perhaps we can send the marbled murrelet into oblivion and the Earth will not be noticeably imbalanced, just significantly less mysterious, but I don't believe it will be so easy with salmon.

At the end of our interview, Bill made a poignant comment.

"We're like the Aztec culture that thought they had to keep sacrificing maidens every day to make the sun come up . . . We think we have to keep cutting down old-growth trees to have jobs."

CHAPTER 67 SALMON AT GRIZZLY CREEK

I HAVE been reading about salmon this week, and yesterday, even though it was raining, I suddenly felt compelled to take David to see if we could find spawning salmon. It is the beginning of December and, according to my research, spawning should still be occurring. I have seen salmon at this time of year swimming up the main stem of the Eel River, but I have never seen them in a small creek.

I called Chuck to see if he had any suggestions for which creeks to try.

"I'll make some calls and get back to you," he said.

After ten minutes he called back and said, "I couldn't reach anyone at Fish and Game, but I called the store out in Carlotta." This meant that he had called Martin and Shirley's Market, a source of information about local conditions on the Van Duzen River. When I first moved to Humboldt County, I was told that if I wanted to take my children for a day of swimming in the sunshine, I could know if the fog had cleared upriver by calling Martin and Shirley's. Now, in the fall, they are a good source of fish information.

"They said there are carcasses," Chuck informed me. "You might try upstream from Pamplin Grove at that little turn-off by Healy Creek, or you could go on up to Grizzly Creek and look there."

Even a chance that I might lay eyes on the carcass of a salmon that had spent a year wandering the Pacific, had found its natal stream along hundreds of miles of coastline, and returned home to spawn was really more than I had expected.

"I wish I could take off work . . ." Chuck stalled. "It sounds like a lot more fun to be looking for salmon than doing what I'm doing . . . but I better not . . ."

David and I had only four hours before I had to teach art. We had an hour drive each way to reach Grizzly Creek, which flows off MAXXAM land, through Grizzly Creek State Park, and then into the Van Duzen.

"Thanks, Chuck," I said, eager to get going. In a way, I wanted to share the adventure with just David, so that we would have to piece the evidence together ourselves.

As we drove, I was tempted to stop and check the creeks on our way south. Elk River flows under the freeway just south of Eureka. I speculated that if we hiked down the railroad tracks to its estuary, we might see salmon entering the river mouth from the bay. But I kept driving . . . Salmon Creek, the next creek south, also comes out of Headwaters . . . there *should* be salmon in a creek named Salmon Creek . . . but there probably aren't anymore. We passed Fernbridge, which spans the Eel. We could hike down under the bridge, past the gauge that in winter measures high water, and perhaps see salmon passing by . . . but we kept driving, turned east onto Highway 36 through Hydesville, past the Mighty Mart; through Carlotta, past Martin and Shirley's. We reached the turnout over the river near Healy Creek, and a man was standing there looking out at the Van Duzen.

"Someone's there," I said.

"Better not stop," David advised, cautiously.

I agreed. The man had a sour expression as he stood staring into the river.

"Maybe I should have interviewed him," I considered aloud after we had passed. "Perhaps his sour expression is simply because he knew the river when salmon were abundant. Maybe he has a story to tell."

"I don't want to get killed looking for salmon," David stated flatly.

I kept on driving, weaving with the highway between the massive trunks of the ancient redwoods of Pamplin Grove County Park. David was doing his homework beside me, and it was fun to be together in a warm, cozy car on a rainy day.

We pulled into the parking lot of Grizzly Creek State Park.

"Get your raincoat, okay?" I asked David.

"Do I have to?" he countered. He is adapted to the fog and cold in a way that I am not. After I insisted on the coat and we started hiking, I was a little embarrassed to notice how warm the day was, despite the rain.

As we reached the creek, the rain let up altogether.

"I wish I hadn't brought this . . ." David said. "There's one!" he suddenly interrupted himself.

I followed his eyes and there, on the opposite shore, was a huge dead salmon. I was shocked. Could it be this easy? The fish was two or three feet long, lying head down at an angle to the waterline. I stood like someone on a pilgrimage, my destination suddenly reached before I had scarcely begun. Meanwhile David was figuring how we could actually cross the muddy water to reach the fish. We walked to the mouth of the creek, where the water spread out over gravel before it entered the Van Duzen. It was still too deep to cross even though we both wore rubber boots.

"Let's go up to the highway bridge," I suggested.

"There's another one!" David said as we headed back. Another salmon, also on the other side.

"There's another one! Funny how they're all on the other side," he said again. I felt like we were participating in history. How much longer would there still be salmon in this silt-laden creek?

We stayed close to the guardrail as we crossed the highway bridge, watching for cars emerging from the shade of the giant trees on the narrow two-lane road. Then we walked downstream again toward the first fish David had spotted. I bent over it, absorbing details into my mind. Its size alone was startling. Though I knew the fish had most likely simply followed ancient inner drives to migrate and spawn, it carried with it even in death the romance of the open ocean. It traced a wide arc through my imagination, following the west coast of North America northward into the Gulf of Alaska and returning, spanning months and miles with the fanning of the tail that lay before us, worn to bare cartilage.

We deduced from the irregular black spots on its back and from its size that the fish was a coho, the species that has just been listed in Northern California as threatened. Coho spend one year in their natal freshwater creek as juveniles and then migrate to the ocean. Salmon evolved sixty million years ago. They are more ancient than 80 percent of the fish alive on the earth today. Two million years ago, in response to the conditions of the Ice Age, they evolved a tolerance for saltwater conditions which allowed them to take advantage of more plentiful food supplies in the ocean during their adult life.

The signals prompting the homeward journey of this salmon may well have been sensed last June, when my children and I were just gearing up for our summer vacation, and day length was just beginning to wane. Within this decomposing carcass, day length cross-referenced with magnetic field to set this battered tail fin fanning in a gradually homeward path. It is believed that once a salmon reaches its home coastline, it follows the shore, picking up the scent of each creek along its way. The smells of the rotted algaes, the soils washed out to sea, even the scent of the specific animals that dwell and die within a watershed, mix and reach the salmon's extremely keen sense of smell as the fish passes by river after river. Eventually, one particular combination of smells draws it into the estuary where, at the beginning of its life, it underwent the remarkable process of becoming acclimatized to life in salt water. After an entire year at sea, individuals of each particular "run" of salmon, that is, those that hatched together in a particular creek, are genetically coded to return to their natal watershed simultaneously so they can spawn en masse.

As soon as the salmon sense that river flow is adequate following autumn rains to provide entrance to their natal creek, the salmon begin moving against the current. Once they enter fresh water, they cannot delay. They eat little or nothing and their body chemistry begins to change, bringing about a dramatic transformation in their physical appearance. A female's body fat not only fuels her trip upriver, but converts into nutrients for the thousands of eggs she carries within her. The fat within the body of each male likewise converts into energy for sperm production. Salmon that once traveled together in schools for protection become more and more aggressive toward one another as they move upstream, and their bodies, particularly those of the males, develop enlarged jaws and teeth, and thickened skin to withstand the attacks of competing salmon.

I looked at the worn skin, the frayed fins and tail of the dead salmon on the shore, and wondered if it had actually found clean gravel and a mate, and if its long journey actually resulted in fertilized eggs securely lodged between stones for the winter. We studied each of the dead fish in turn and then headed back upstream toward the highway bridge.

David watched the shore while I watched the muddy water of the creek intently. Suddenly, I realized that I was looking at water spilling sleekly off the back of a living salmon! This fact was confirmed by the appearance of a tail fin sticking partway out of the water.

"Look! There's one that's alive!" I yelled to David, grabbing his shoulder so I could direct his line of vision down my arm into the nearly opaque creek.

"There's another one!" he said, his sharp eyes now attuned to life.

The salmon were working their way upstream, and as we watched, one snaked and flopped over a shallow riffle and disappeared into the next pool. As we focused into the deep water, we saw that four salmon were already in the pool. In the silt-laden water, salmon moved up alongside one another into pairs, hung in the current together, and then drifted apart. One pair began folding over and over one another, and I realized that they were doing something I had just read about. During courtship the male salmon brushes the female with his body, caressing her with his fins so that even at night, or in water where visibility is low, the female can know he is a male by the feel of his large, stiff fins. The scene became dreamlike. Salmon are so much a part of American mythology that it is easy to forget that they are real. I felt as if I had suddenly spotted the Pilgrims landing on Plymouth Rock. I was watching salmon! David was seeing a once commonplace sight that few people he will ever meet will have seen.

CHAPTER 68 LUCK

AS I watch the salmon, I recall my conversation about Gaia on the telephone with Jesse Noell almost a year ago. His voice falters as it takes time to reach back millions of years. "The salmon were here before the land, before the forest. They have been here sixty million years. They swam over Headwaters when it was still under the ocean." He speaks like a patient teacher, as if he were in the yet unnamed Pacific sixty million years ago and were merely recounting. He is one of those people who are more fish than human, and I can see him powering with the side-to-side movement of his body, finding by scent his natal creek, feeling the magnetism of the land, drawn up a watercourse that is now miles inland.

"The male waits, suspended in water as the female lifts the gravel with the suction that follows the repeated lifting of her tail. For sixty million years, the salmon have been cleaning these streams year-round, run after

run, fall-run salmon, winter, spring . . . " Cleaning the streams? Who but Jesse would think of the salmon cleaning the streams? Who thinks about cleaning a house in which he does not live?

"Gaia is the opposite of entropy," Jesse declares quietly.

"What is entropy, again?" I ask.

"Entropy is a law of thermodynamics that describes how systems are constantly wearing down and becoming disorganized. It is easier, for instance, to combust fossil fuel than to try to recreate it after it has been combusted." I thought about how easily I drive and burn gas. The idea of recreating even a cup of that gasoline opened my mind into a river of time.

"Gaia is the boot-strapping of ecosystems up from nothing. The way our culture defines the world is through entropy, but the reality is Gaia." I wondered, is this how Jesse spends day after day on the telephone, going to meetings, because he has this fundamental optimism that life is always coming into its own?

"The salmon have brought the nutrients from the ocean, have died and fertilized, with the decomposition of their own bodies, the creek where their eggs will grow. Their bodies have been dragged inland and the forest floor fertilized. The redwoods are an expression of that force, drawn from the wide Pacific, pushing upward hundreds of feet, processing and transpiring hundreds of gallons of water a day, raking fog from the air with their needles and raining it back down."

To flow as Grizzly Creek, Elk River, Freshwater Creek . . . The longer I have worked on this book, the less "entropic" my life has seemed. Chance coincidences that no longer seem chance, information that appears to all but write itself into the book, have made me feel that the forest has been writing its own story through me. But perhaps it is Gaia, seizing us gently like a mountain lion picks up her kittens, and padding with us softly toward enlightenment.

The female salmon suddenly turned sideways at David's and my feet and flapped her tail vigorously.

"This is how she cleans the gravel of silt and digs a nest!" I whispered.

"This digging may go on a long time, making a deep hollow in the gravel. It's what attracts the males."

I watched for the female to begin probing with her tail, a signal that she might be about to lay her eggs. At the same time, I knew I was being a little unrealistic. The streambed was so coated with silt that the effect of her tail flapping was inconsequential compared to the amount of silt that remained undisturbed.

"For the female to lay her eggs, she goes into a crouch by actually opening her mouth and taking water into it so that the current forces her tail deeper into the hole." I thought of her gills and wondered how much silt they could tolerate. On the other hand, I reasoned that these fish were two years old. They were evidence of at least a slim survival of this creek's run. What a precious thing this single run of salmon is, I thought, genetically molded to withstand normal erosion of these highly unstable slopes, its internal clock tuned to this one unique place in the world though the fish themselves range over thousands of miles.

After I got home, I told Chuck about the salmon.

"If it's this easy to see salmon, why haven't I seen them years ago?" I was honestly perplexed. I spend a lot of time along creeks, but had I just never looked? How could I miss the carcasses?

He smiled.

"You lucked out. Why am I not surprised? I have only seen salmon like that twice in fourteen years that I have gone out looking. My boys and I have stood staring into creek after creek and seen nothing. Suddenly, on a rainy day, you say, 'I'm going looking for salmon,' and there they are."

Poison oak

CHAPTER 69 PEPPER SPRAY

THE world has been stunned by the release of video footage of Humboldt County law enforcement officers swabbing concentrated pepper spray directly into the eyes of nonviolent demonstrators, several of whom were teenaged girls. Will this book ever end, or will I be like the painter Francis Bacon, who worked on paintings so long that his agents used to walk into his studio and take each painting away as they deemed it finished, leaving a blank canvas in its place? My deadline approaches. Will I have to break off midsentence because my editor has flown out here and said, "This is it. I don't care what else happens. The issue may not be over, but the book is done"?

My sister, who lives in France, saw portions of the video footage of the pepper spraying on her television. She called me and asked, "What is going *on* over there?"

I tried to describe the political climate in which a police force would actually film itself, for the sake of future instruction, inflicting lawless torture on passive citizens. By the time I was finished, however, I think she had more questions than before she called.

So I went down to the law office of Mark Harris, who has again joined forces with attorney Macon Cowles, this time to represent the nine demonstrators who were pepper-sprayed. The attorneys are alleging that their clients were subjected to unreasonable use of force, excessive detention, and were denied their basic rights of free speech, association, and to petition government under the First, Fourth, and Fourteenth amendments to the Constitution.

Mark lent me copies of the police videos taken during demonstrations at the Pacific Lumber Company headquarters in Scotia on September 25, 1997, and at Congressman Frank Riggs's Eureka office on October 16. I have sat for hours transcribing the videos word by word. As I have worked, rewinding the tape over and over to clarify a garbled sentence or disentangle screams, seeing the same bodies recoil time after time, I have kept in mind not only my sister's words, but those of a friend's fifteen-year-old son who said, "They could have just left. They didn't have to be there. They were breaking the law." I have silently countered that this could be said of the black civil rights demonstrators in the South who forced their way into the front seats of buses, into restaurants, or into line at drinking fountains. It could also be said of the American colonists who unloaded British tea into Boston Harbor.

The following is my complete transcription of the September 25 video filmed in Scotia.

The cameraman pans a rally under way outside of Pacific Lumber Company headquarters. The focus is the preservation of the entire 60,000-acre Headwaters Forest ecosystem, not just the 7,500 acres allotted in the bill which was approved by Congress and signed by Clinton a month before. Demonstrators are chanting, "Hurwitz out of Humboldt. No Deal! . . . Hurwitz out of Humboldt. No Deal!" and "PL, yes! MAXXAM, no! . . . PL, yes! MAXXAM, no!"

Next the cameraman moves indoors and focuses on four young women and three young men who are sitting cross-legged on an uncarpeted floor of perhaps marble or linoleum, their arms linked within twenty-five-pound, black, V-shaped pipes known as Black Bears. They have their heads bowed and they are softly singing, "Keep your eyes on the prize. Hold on, hold on . . . Keep your eyes on the prize. Hold on . . ."

One female demonstrator states in a soft voice, head still bowed, "Civil disobedience in this country goes back a long ways . . ." The rest of her sentence is lost on the tape. Without lifting their heads, the seven demonstrators then begin to chant, "Sixty thousand acres, no compromise . . . Sixty thousand acres no compromise . . . Not one more ancient tree . . . Not one more ancient tree . . ."

A sheriff interrupts the chanting, by introducing himself and then delivers his prepared message. Not all of the words are audible on the tape.

" . . . Sheriff Delaney . . . Your conduct here is in penal code violation of section 407, unlawful assemble, and 602, trespass. I command you in the name of the people of the state of California to disperse. If you do not, you shall be arrested for violation of section 407, unlawful assembly, and penal code section 409 and 416, refusal to disperse. If you resist, you will be charged with penal code section 148, resisting arrest."

The demonstrators continue to sit with their heads bowed. Standing around the demonstrators, out of sight of the camera, are ten to fifteen male police officers, according to a recent interview with Mark.

"I am requesting that you leave immediately within five minutes or you will be subject to arrest. It is my intention if you do not leave to use pepper spray or chemical mace to extricate you from your . . . from those steel cases. You have five minutes."

"Sir?" one of the demonstrators begins. "It is only legal in the United States to use pepper spray in the event that you feel threatened."

"I understand. Folks, you have five minutes."

The demonstrators' words are softly stated and intermingle so that only fragments such as " . . . threatened . . . nonviolent . . . and that's a lawsuit . . ." can be heard as they attempt to reason with the police.

"People with asthma can die from pepper spray," one girl states.

"I have severe asthma. I cannot be sprayed," another demonstrator with blond curly hair impassively informs the officers.

"Then release yourself, lady."

"I am a public school teacher in Humboldt . . ." She continues.

Then a coarse voice from beyond the camera's range threatens, "You've got an education coming."

My body is growing increasingly tense. At this point I recall my friend's son asking, "Why don't they just leave?" I picture the demonstrators slipping their hands from the sleeves, standing up, straightening their clothes, looking the officers in the eyes, and saying, "Oh well, never mind. If you're going to torture us, forget it. We're leaving. This is too much trouble."

The officer continues, "Folks, listen up. It is now two minutes to two o'clock by my watch. I gave the dispersal order three minutes ago by my watch. I am asking you again to please remove your hands from the metal pipes and leave, or, again, we are going to use chemical spray or pepper spray to try to get you to remove your binds. Will you please vacate these premises? You're trespassing."

No one moves. Heads are bowed.

"Okay. It is my intention to use pepper spray or chemical mace to extricate you from your binds. . . . Two minutes, folks."

"Okay, folks, your time is up. I've given you five minutes. Are you going to leave your binds and leave this building?"

No one moves.

"Okay, we'll be starting using the pepper spray then. Marvin . . ." He says this all very matter-of-factly, almost cheerfully, as if the demonstrators are being given instructions prior to the submarine ride at Disneyland. I commented to a friend of mine that at least the officer was polite. That person looked at me patiently and said, "That's no excuse."

The officer councils, "Do not resist. If there is any resistance to us, any threats on us, it will be a felony assault on a police officer . . ."

"We're nonviolent . . . we're nonviolent . . . we're nonviolent . . . we're nonviolent . . . we're nonviolent . . ." the demonstrators state over and over softly.

"I suggest you leave. You've made your point."

There are muffled comments from the protesters. One hears the word, " . . . nonviolent . . ." again.

"Let yourself go now," the same officer says in his business-as-usual tone.

On the table behind one of the demonstrators, the viewer can see a sheriff set down a paper cup and dip a swab into it. He advances behind the closest young man, pulls back his head, and liberally swabs his eyes as if he is spreading peanut butter on bread. After a moment the pain sets in, and the young man begins a deep and private moaning. One feels that the moaning would be high-pitched if he had not constricted his entire being.

From off camera, one hears a female demonstrator shout, "No!" as if the ancient matriarch within her has seen enough and decided to take charge of the scene.

"We're with you, brother," another demonstrator consoles from off-camera.

"Let go and we'll give you water for your eyes," one of the officers bribes the young man in pain, while the next young woman is asked, "Are you going to let go?" He then pulls back her head and spreads on the pepper spray. She begins screaming, which sets up a cacophony of protest from the group attempting to be heard by the public outside.

"These people are being hurt!"

"Pepper spray! Pepper spray! Pepper spray!" But only distant voices of the rally in progress can be heard over a loudspeaker.

"Pepper spray! Pepper spray! There's no need . . ."

"Ma'am, we're going to use mace next. It's worse."

The young man next in line to receive the pepper spray says that he will release.

Meanwhile the officer has moved on to a girl I know who lives here in Arcata named Vernell, or Spring. She is sixteen years old. The officer with the pepper spray places his hands on her shoulders.

"Ma'am, are you going to release?"

The dispassionate cordiality of the officers is wearing me down. I wish with all my heart that the local media presented a more balanced and comprehensive picture of this issue, so that at least one of the officers might suddenly exclaim, "Stop! This is morally and legally wrong!" These policemen have probably not seen the enormous clearcuts on MAXXAM land. They very likely do not know the crimes that are being committed just beyond the front ridges. I wonder if they have seen photographs of the ghost towns left after industrial forestry has liquidated an area.

I begin to cry . . . for Vernell . . . and for our community turned against itself. Hurwitz, the governor of California, the head of the Board of Forestry, the directors of CDF, do not use their own hands to hold Vernell's shoulders. They do not seize her head and hold it tightly while she struggles. They do not apply the concentrated capsicum to her tear ducts . . . I am appalled at how power is being used to intimidate a sixteen-year-old girl who is not threatening murder, who has stolen nothing, who simply confronts a system that is unwilling to listen its own people and does not have the wisdom to protect its own forests. I watch the officers and I answer my friend's young son, "They cannot leave because they are the trees who cannot choose new forests. They are the salamanders who cannot hike to new watersheds. They are the marbled murrelets circling in search of trees that no longer exist. They are the salmon who return home to creeks filled with mud. They are the salmon fishermen who have invested their lives in an industry that has been sacrificed without their permission. They are Kristi Wrigley, whose apple orchard is rooted where it was planted a hundred years ago. They are Mike O'Neal, who cannot sell his house because it may be swept away this winter. They are ourselves."

"Humboldt County, log it or leave it," I saw yesterday on the metal frame around a license plate. Did anyone teach the owner of that truck or these police officers about the beauty of sustainable forestry so that they would be too well informed, too wise, to commit such acts of terrorism, even though it might be ordered by their superiors?

Spring sits silently. The officer firms up his grip on her, secures her head, forcibly tips back her face, and puts the pepper spray in one eye. She recoils and begins screaming, "My eye! My eye!"

"Are you releasing?" Now three officers hold her down. She pulls an arm out of the metal sleeve, and they back off for a moment.

"Don't let her stick it back in there!"

The three men twist her arm behind her back and forcibly pull her head back again.

"Are you releasing?"

Meanwhile the teacher is yelling, "Torture! Torture in the United States! This is a goddamned death squad! Stop it! Pepper spray is a lethal force. We do not need it. We never needed it."

"I'm gonna do the other eye if you don't release," the officer with the pepper spray threatens.

"She is your daughter!" the teacher screams.

"This is your last chance. Are you going to release?"

One officer gets a better hold on Spring's head while the other puts pepper spray in her other eye. The next girl in the circle, a girl with pigtails who looks even younger than Spring, watches and then lowers her own face, crying.

The teacher continues to scream. "This is an outrage! Goddamn it! Stop it now! Don't you have a conscience?" It is she who has asthma, and at that point she agrees to release. Officers handcuff her as others move in on the girl with pigtails.

"Please don't hurt me. I don't want to be hurt."

"Then let go of the pipe."

"She's in solidarity with her friend!" the teacher screams as she is being taken into custody. They pull the girl's head back and swab both her eyes at once. She sits absolutely silently, head up, eyes closed, utterly silent.

Then they return to Spring.

"Ma'am, we're gonna put some more on," they threaten. "You've got to release." Then they tackle her even though she is curled up on the floor in pain.

"No! No!" the teacher yells. "She is our mother. She is our daughter. She is our friend . . . Stop it! Stop it! Oh my God!"

They reswab both of Spring's eyes.

From beyond the view of the police camera there comes the sound of male laughter.

The scene gets increasingly chaotic. The camera moves around in a dizzying scene as the officers reswab the already blinded eyes of the demonstrators at random. Only four demonstrators out of seven are left. They are slumped over or lying on the floor, three girls and one young man, all with eyes closed. Each of the demonstrators has been swabbed at least twice in each eye. They are completely incapacitated.

Yet, even though pepper spray has been deployed far in excess of manufacturer guidelines, officers have not succeeded in forcing these four demonstrators to abandon their protest. In an abrupt change, an officer approaches the girl with pigtails holding a squirt bottle containing water.

"Ma'am, water . . . this is water. Open your eyes!"

The girl panics.

"How can I trust you?" she asks in a terrified voice.

"Trust me, it's water."

"Would you do this to your daughter?" she asks, crying, her eyes still closed.

The act of spraying water directly into someone's face exhibits a level of rudeness by which to gauge the magnitude of the brutality I have just been studying. Even though the spraying of water eventually seems to bring some relief, the sudden change to concern by an officer is strange. Eventually the police follow normal procedure for lockdowns, carrying Spring and the girl with pigtails outside of the building on linked stretchers. Asbestos blankets are laid over their bodies and in four and a half minutes the weld on the sleeve of the Black Bear is chipped away using a metal grinder. Then a chisel is used to pry the sleeve open. Meanwhile the other two demonstrators have been carried out, and they release themselves from their Black Bear. Then all four demonstrators are handcuffed and taken to jail.

I think of my friend's son's question, "Couldn't they have just left?" and I counter, "Can't MAXXAM just sustainably manage its forests? Can't it just cease salvage logging and harmful winter operations? Can't it just think of community and global well-being?

Pepper spray was used on demonstrators again on October 3 out-of-doors at Bear Creek, where two men were swabbed, sprayed at close range, and eventually arrested.

The October 16 incident was similar to Scotia, in that it also occurred indoors where fire danger was cited as justification for the use of pepper spray, even though asbestos blankets are laid down to catch any sparks. The demonstration began when Earth First! activists moved a large stump into pro-timber Congressman Frank Riggs's office, and four girls entered, said good morning to Riggs's staff, sat down around the stump, and linked arms within the same type of heavy, black pipe as those used in Scotia. Sawdust was sprinkled on the floor, while one of the staff people continued to talk on the phone. The other activated an alarm and then walked to the door, stating, "No one is leaving." Demonstrators who brought in the stump left out a side door, saying, "Good luck. We love you," to the four girls seated around the stump.

A similar scenario to Scotia ensued, with ten to fifteen male officers taking part or standing by, and the demonstrators warned and then swabbed. One woman was swabbed seven times on one eyelid, and another, after all but two women were released, was sprayed directly, with the spray nozzle held just two inches away from her eyes, which were being forced open.

In an interview with the *North Coast Journal*, Mark Harris, who began defending demonstrators in 1990, when four young men had their heads forcibly shaved in the county jail in Eureka, said, "Civil disobedience on the North Coast is very different [from] civil disobedience on college campuses or in urban areas. There's no one . . . to see [police misconduct]."

He said, "Harassment . . . takes the form of everything from the liberal use of very severe pain compliance holds on nonviolent demonstrators to excessive detention pretrial—putting kids in jail a long time without even an arraignment, dragging them through pretrials only to drop their case, chalking up a bunch of charges which you ultimately dismiss, using our judicial system short of trial to hurt people . . . because they are politically opposed to your point of view."

Referring to instances such as the November 15, 1996, rally at Fisher Gate, Mark described the willingness of law enforcement officers to declare a gathering an "unlawful assembly."

"This is often the first thing out of their mouths. But to make that call, you have to show that there is some imminent peril, some danger to the public or to property or a riot about to occur.

"We need to compare the amount of force used in all its manifestations including going to jail . . . to the harm caused. That is the essence of a Fourth Amendment analysis."

Varied thrush

CHAPTER 70 SPRING

I ARRANGED to interview Spring, the sixteen-year-old girl who was pepper-sprayed at the Pacific Lumber Company headquarters. For the past two weeks, the local media has hammered away at us from myriad angles trying to convince us of the illegitimacy of Earth First! We have been told that Earth First!ers are not interested in real dialogue; that they are just showoffs; that the pepper spray didn't really hurt; that the screaming was staged; that they're outsiders; that they're dangerous. And, with regard to Spring and a couple of the other girls, we were indignantly asked why these minors weren't in school.

Before the day of the interview, however, news of another incident surfaced that was very interesting relative to the pepper spraying. A CDF inspector had found on MAXXAM land illegal road failures and overcutting of such magnitude that he stuck his neck out and filed two citations with the county district attorney. MAXXAM was currently on probation for previous violations. These new citations were violation of that probation. Incredibly, however, once MAXXAM was notified by the county DA that charges had been filed, company heads went to the inspector's superiors, who "requested" that the CDF inspector withdraw his citations from the record, which he did. It was with this in mind that I kept my appointment with Spring.

As I walked into the cafe to meet Spring, I spotted her bent over her backpack, searching for her wallet. She stood up to greet me, and I was surprised to see that she is as tall as I am, about five foot nine. Her slim body radiates physical strength in the carriage of her shoulders and the way she moves. As we ordered, I noticed that she is very polite and careful in her choice of words. After we sat down, I began to study the clarity of her eyes, the delicate, clear delineation of the lids and the lashes. As an artist I study faces for how I might draw them, but as a mother and a believer that we are what we eat, I study them for their health. As I admired Spring's beauty and unpretentious intelligence, I became aware that her tongue was pierced. A gold bead moved almost out of sight on the surface of her tongue while she told me about her life.

She told me that she became aware of the Headwaters issue when she was eleven and moved from Oregon to Arcata.

"My father became involved and I learned about it from him."

"Did you know that you would become this involved yourself?" I asked, still growing accustomed to the bead.

"No. All of my life I have been a really persevering student, and I was busy with my schoolwork."

I smiled at the neatness of the word "persevering." Spring's expression was serene and informative. This was hardly the language of a terrorist.

"I became involved with the forests in Oregon, but I have a special feeling for Headwaters because it is part of my home bioregion."

Again I smiled to myself, enjoying the way she positioned herself globally, at the age of seventeen.

"Have you been to Headwaters?" I asked her.

"Oh, yes," she said, politely not making my question look ridiculous by assuming that I was referring to the Headwaters Grove specifically. "But I spend most of my time in the more vulnerable groves. Those are the ones I really know well."

"You mean the ones that are not part of the Deal—Shaw, Allen Creek, Owl, All . . ."

"Yeah, I haven't been in Headwaters or Elk Head Springs as much. I love Shaw."

I looked at her with sudden recognition. The salamander within me suddenly glimpsed its own kind across the table.

"Yeah, I love Shaw too," I said to her, quietly.

"It's so wild," she went on. "But the press just wants to talk about pepper spray." She thrust an imaginary microphone up to my chin, "'Yeah, but did it hurt?' they ask, while I am just trying to draw attention to the forest issues."

"But perhaps this is what it takes to get people to notice that something is really wrong out here," I offered.

"What I really like to do is tabling and community outreach," she said. "I like to do bake sales. People are standing there eating, and we start talking, and then they see who I really am. It is through one-on-one conversation that people come to understand the issue. Almost everyone I know has become involved with Headwaters because of someone they know."

I could picture a shopper in Fortuna munching on a whole wheat brownie, regarding Spring as if she were from outer space, but from a planet that might be a surprisingly nice place to visit.

"I love meeting people," she went on. "I have a lot of connections with the lumber workers in tree sits in the woods. At a blockade"—and by this she meant a point on a logging road where the passage of trucks has been temporarily halted by Earth First!ers chained to a bulldozer or suspended from a tall tripod made of logs, for instance—"we have coffee and doughnuts for the workers, and we hand out copies of the Stewardship Plan. We've gotten to know Climber Dan."

"Who is Climber Dan?" I asked.

"He's a tree-topper who climbs really fast. He has to do his job trying to get us down, but he knows what's going on. He doesn't want them cutting down all the forests.

"In the tree village in Owl Creek, he's the one who cut me down."

"How long did you tree-sit?"

"Two weeks, there."

I can count on one hand the number of times that I have spent two weeks in any forest on the ground. What was it like for this girl to spend two weeks high in a redwood? I wondered.

"The forest was so beautiful when we got there. Slowly, each day, more trees would fall. It got hotter and drier, and there were fewer and fewer animals. The animals would come out at night after the loggers had left and the Cats had stopped, and when the workers would arrive the next morning the animals would start shrieking. Then finally it was like a parking lot."

How little we know about one another. How surprised the person would be who was quoted in the newspaper asking why Spring wasn't under the supervision of her "parent or guardian," to learn that she had spent two weeks on a wooden platform suspended from a redwood. I cringed, imagining how distraught I would be if my own child were doing the same thing, but I no longer had the heart to ask this young woman, "Wouldn't it be better if you tried to work within the system?" If the system is not working for a CDF employee who is prevented by the state from performing his job, how could it work meaningfully for a sixteen-year-old girl?

I recalled the meeting hosted by the residents of Freshwater at their elementary school with the very superiors who likely pressured the removal of the MAXXAM citations, when the people of Elk River and Freshwater Creek were told that "science" has not "conclusively proved that clear-cutting has a negative impact on a watershed." And that CDF would "consider" points made, and that "studies are under way," and that "this is just how steep, unstable slopes behave." Meanwhile the cutting goes on. Spring is essentially, in her own way, trying to compensate for the inspectors' inability to be effective. She is trying to act in the place of a district attorney who has withdrawn charges. The "system" is working for MAXXAM, but we have sixteen-year-olds dangling on boards trying to make it work for us.

"Every time that Climber Dan would come up, I'd unclip all my safeties, and that just freaked him out because he was afraid if he tried to get me I'd fall. There were two of us on this narrow door, and it was starting to crack. They cut a tree that was ten feet away from us and then bulldozed the stump so there'd be no evidence. Bulldozers were ramming and sideswiping the tree that Coho and Sawyer were in. There was a beautiful grandmother tree that we were trying to keep them from cutting."

I envisioned my grandmother tree in Elk Head Springs.

"Raindrop and I were up on the platform. I've learned a lot about falling trees from watching the logging. They began to cut the grandmother tree and I looked down and saw that they were preparing to cut it straight toward us. They were bulldozing the fall bed right to the base of our tree. I thought we were going to die. We were hugging each other, and I thought, 'Well, this is a good way to die.' The tree fell, and it's top was just twenty feet from our tree."

I imagined the force of the wind and the vibration of a three-hundred-foot tree's fall, not to mention the emotional impact of seeing one of the great giants felled.

"How did it end?" I asked. "Why did you come down?"

"One night one of the sitters decided we weren't in the best location to stop the cutting. I was so tired, but I agreed to help move gear so people could switch their trees. We just kept moving gear and in between I'd sleep. It was about five o'clock in the morning when I heard, 'You're under arrest!' I woke up and it was starting to get light. The workers were starting to arrive, and they had caught some of the people sleeping on the ground."

"Where were you?" I asked.

"Oh, I was in my tree. I tried never to be caught on the ground. But I had lost my ground support. Climber Dan came up and lowered me down with a rope. It was real dangerous. If I had put my arms up I would have slid out of the noose. I felt it sliding up toward my throat."

I could feel the interview drawing to a close, so I finally had to ask, "How do you justify for the press the fact that you were trespassing when you entered PL headquarters and locked down?"

"We were engaging in nonviolent civil disobedience. Civil disobedience has a long history of effective use in this country. By going into MAXXAM headquarters, we were trying to draw attention to the Office of Thrift Supervision hearing. There was a mock trial going on outside, but Carl Anderson, the security guard for MAXXAM, put butcher paper up on the windows so people couldn't see us."

The OTS hearing is an investigation into the liability of Charles Hurwitz with regard to the failure of United Savings Association of Texas, which was controlled by United Financial Group of which Hurwitz was chairman of the board and a 23 percent owner. The bailout by the federal government of USAT was the third most expensive in American history.

"Did you know that they might use pepper spray to stop you?"

"I knew that Officer McAllister used it on Crow in the tripod at Bear Creek. But we didn't know about their policy to use it routinely. They were trying it out, that's why they were filming it. They were making a training film, so we were trying to show that pepper spray wouldn't be effective by hiding how much it hurt. Their insurance company was videotaping it too."

"Their insurance company?" I asked incredulously.

"Yeah. They said it was fine if they used it, just to wait until they got there before they did."

"It looked like there was one policeman who had concern for you," I said.

She made a face. "Yeah, that's what I thought, too. Then I saw the videos and saw that he was the one who had applied the pepper spray. It made me feel sick. You know, 'Let me hold you while I beat you.' I was blinded, but my hearing was really clear. It was awful to hear them hold your friends down and then hear them start screaming."

I could see Spring was tired. She had a cold, and she said that she had gone several times to San Francisco for press conferences. I thanked her and cleared both our plates, a little out of maternal protectiveness, but more out of humble respect for this remarkable young woman. I thanked her and we said goodbye.

I spoke to Jesse later and asked him if he thinks Earth First! is a help or a hindrance in the issue of Headwaters.

"Everyone has a role to play. I wouldn't want to say what will work and what won't," he answered.

I have been helping David with American history lately. I asked myself, could America have won its independence simply through discussions with the king of England? Or would we have been told at best, "This requires more study," "We'll look into it," and "There still isn't adequate science"?

CHAPTER 71 CHILDREN

FROM the age of twenty-seven to forty I raised my stepson on our farm in Vermont. His name is Jesse and he is now twenty-six, almost the age I was when I began raising him. When Jesse was just ten, he could tell a hay bale that might internally combust and set our barn on fire from a fully cured bale, just by hefting it. He could hand-thin long rows of carrots efficiently. He could do the chores alone, feeding and watering hundreds of animals and not miss anybody. When we are children, if we are lucky, we learn our parents' work. Jesse is an organic farmer by birth.

Now David, after working with me for the past year, can tell you if a stream is healthy and why. He can tell you if a road will fail and where it needs water bars. He knows how to tell old-growth on a ridge from second-growth. He knows what a forest looks like that has been thinned to efficiently release growth in the remaining trees, compared to one that has been simply plundered. He knows that CDF stands for the California Department of Forestry and that it answers to the Board of Forestry; that Fish and Game is a state agency; and that Fish and Wildlife is federal. He knows how to walk seventeen miles in two days, not only without complaining, but with joy.

But I worry about what I have done by making David knowledgeable about systems that may fail him. My stepson is still working the farm where he grew up. But what is the future of the forests here? What is the future of agencies like CDF, that are supposed to represent *us*, the people of California, in the stewardship of our forests? What is the future of the Endangered Species Act itself? Will it continue to exist as a valuable tool for protection or will it be swallowed into pages and pages of Habitat Conservation Plans while our precious native species are horse-traded into extinction?

Salmonberry

CHAPTER 72 FÜR ELISE

I SIT out in the backyard with David. It is night. He is reading and I am staring up into the trees. I have been watching the pepper-spray videos. This is my job, to report, but it takes its toll on my heart. David is so cute, his thick book turned so the pages catch the porch light. He is supposed to be in bed, but he has finagled a few extra minutes of reading. Those minutes stretch as I am soothed by the night air. The moon is bright, and clouds are moving overhead. In this foggy climate, genuine clouds and stars are rarities. I savor the breeze and the clarity of the night.

Then, inside the house, Suzanne begins playing the piano. She does not like to play in recitals, but I am blessed with these impromptu performances, heard distantly, when she does not even know I am listening. She has almost learned "Für Elise" by Beethoven. I have heard her practice bits of it over and over, but now she plays the whole piece with new feeling. This is new for me, to hear her play Beethoven, and to hear her play with feeling.

David reads silently, and as I listen, the word "struggle" comes to mind. Struggle. The other day Suzanne gave David an offhanded compliment about the amount he had been practicing the piano. She said, "I can't take any more of those happy little songs! They are driving me crazy! Do you have to play the piano all the time?" Yet he is playing the same pieces she just left behind. Now, at the age of thirteen, she is ready to play more complex music, and she is cultivating attitudes about struggle through listening to Beethoven emerge from the movement of her own fingers. In the process, she gives me companionship without even knowing.

My heart is sick that communication over Headwaters has become degraded to the point that uniformed police officers are holding the heads of sixteen-year-old girls in the vice grips of their own hands and painting concentrated red pepper into the girls' eyes. If A. S. Murphy, past CEO of Pacific Lumber, who instituted the standards of sustainable forestry and no clear-cutting, could look up from his grave . . . Where is he buried? I wonder. Fortuna? I would like to go there and ask him if he ever dreamed that anyone would be torturing young girls in order to silence protest over the clear-cutting of the forest he never dreamed would be virtually stolen from his widow and heirs.

Is civil disobedience effective? Sometimes. Is it in this instance? I don't know. I am not the kind of person who enters a congressman's office with a tree stump, locks down, and refuses to leave. Yet all the other means of communication—mediation, lawsuits, religious forums, direct one-on-one communication, candlelight vigils, massive rallies—have not stopped the cutting that continues, day after day after day. For twelve years, since the takeover, the citizens of the United States—not "environmentalists," but we, the citizens, people who breathe air and drink water and eat food and are therefore dependent on "the environment"—have tried to effect change within the system. What is an "environmentalist" but simply a citizen who has shed denial, who has opened his or her eyes and said, "It does *matter. Nature does not have an infinite capacity to heal herself, himself, itself. . . . I am responsible. . . . " It is only the media, working as a tool of greed, that uses the label "environmentalist" to somehow discredit a person who is simply a citizen.*

My daughter is thirteen. Spring, the youngest of the demonstrators, was only sixteen at the time of the demonstration. How the cutting of the forests looks to a judge who is fifty or sixty, and has raised his children, or how it looks to an officer who is thirty and has benefits and a retirement fund established through the police force, may be different from how it looks to a young girl with most of her life still ahead of her. . . . It is different from how it looks to a salamander . . . .

As I sit looking up at the trees, I realize that fog has suddenly erased the stars. So many mornings I wake up with a heaviness in my heart that sneaks in just like this fog. In the morning I attempt to chase it away, just as sunshine disperses the fog from the groves. But the next morning it is back. I have read all the little books that teach the tactics: I reason with myself, "You have so much to be thankful for." I count my blessings. I play through worst-case scenarios. But then, when the heaviness appears yet another morning, I finally have to ask my heart, like a child, "What? Why are you sad? This is real, isn't it?" And my heart answers, "Yes." My heart tells me that it would be wrong to simply count my own personal blessings and ignore the political climate around me. This is not a time to cultivate the skills of positive thinking. In this case, those skills would lead to the cultivation of denial. I would be ignoring one of our greatest challenges as adults: determining when to persist in soothing our hearts, and when we are obligated to open our eyes wider and act. One corporation entered this community, took over the reins, hired more workers than the forest can support, began liquidation, and turned us against ourselves. My heart is sick for the young people who sat in a circle, locked down, heads bowed, and it is sick for the officers who have become so inexcusably brutal.

CHAPTER 73 DEMOCRACY

DAVID is studying the birth of democracy. I help him with his homework in the evenings. At the age of ten, he is trying to comprehend the words of the Declaration of Independence. I dissect it for him, translate, use my voice to emphasize meaning, while the words of our local newspaper, the *Times-Standard,* and also the *San Francisco Chronicle,* simultaneously wash through my mind, mixing with thoughts penned by quill in 1776.

. . . *We hold these truths to be self-evident, that all Men are created equal, that they are endowed by their Creator with certain unalienable Rights* . . . The FBI has begun a probe into whether the civil rights of anti-logging protesters were violated when concentrated pepper spray was swabbed into their eyes by Humboldt County authorities. The investigation was triggered by the release Wednesday of a dramatic videotape showing Humboldt County sheriff's deputies yanking back protesters' heads, lifting their eyelids, and then rubbing liquid concentrate on or near their eyes. *Prudence, indeed, will dictate that Governments long established should not be changed for light and transient Causes; and accordingly all Experience hath shewn, that Mankind are more disposed to suffer, while Evils are sufferable, than to right themselves by abolishing the Forms to which they are accustomed.* The videotape also shows authorities spraying the protesters' eyes from just inches away. . . . In one scene, the liquid was sprayed into a chained woman's face. . . . Rodgers [employee of Frank Riggs] added that one of the demonstrators urinated on the office rug after being sprayed. "The officers were wonderful. . . . They were extraordinarily kind." *The history of the present King of Great-Britain is a History of repeated Injuries and Usurpations, all having in direct Object the Establishment of an absolute Tyranny over these States.* [Congressman Riggs] said that by covering anti-logging demonstrations, news organizations glorify "criminal activity by an extremist fringe that seeks nothing less than the complete destruction of our community and way of life." *He has refused his Assent to Laws, the most wholesome and necessary for the public Good. . . . He has erected a Multitude of new Offices, and sent hither Swarms of Officers to harrass our People, and eat out their Substance. . . .* "This is despicable," a retired Los Angeles Police Department officer told one reporter. "This is more revolting than the Rodney King incident." *He has combined with others to subject us to a Jurisdiction foreign to our Constitution, and unacknowledged by our Laws; giving his Assent to their Acts of pretended Legislation: For depriving us, in many Cases, of the Benefits of Trial by Jury.* " . . . that sounded like Nazi Germany," a retired Humboldt County Department of Corrections employee said. *He has abdicated Government here, by declaring us out of his Protection and waging War against us.* A college professor from Tennessee said he wanted to write the protesters and commend them for their "stoic efforts" in the face of police brutality. *He has plundered our Seas, ravaged our Coasts, burnt our towns, and destroyed the Lives of our People.* "A Palm Springs woman called to say she was ashamed to be a Californian or even an American. 'I just can't believe I saw that in America,' she said. 'The police treated teens worse than criminals for engaging in a peaceful protest,' she said. 'I keep telling my grandchildren to get involved . . . I think that's what America's about,' she said, and started to cry."

I think to myself, "Good for you to call the paper and complain. Better still, don't leave protesting to your grandchildren. Inform yourself and tell the people of Palm Springs about the liquidation of our last ancient forests." Realize that one person can make a difference and that there would be no America without a fight. If everyone who was "too old" or had a job or had children or was afraid abstained from the struggle for democracy, we might still be a colony of England. Don't leave this fight to your grandchildren. Dignify it with your presence, with the weight of your wisdom acquired over many years . . . Be the wise elder you are . . . tell us . . . tell America.

Gray fox

CHAPTER 74 RETURN TO OWL CREEK

DOUG called and said, "Have you heard? They're starting to cut Owl Creek again."

Owl Creek—scene of the tree village, the Thanksgiving Massacre, the lawsuit *Marbled Murrelet and EPIC* v. *Pacific Lumber*, which was upheld in the United States Supreme Court, the grove above which I lay in my sleeping bag at the foot of Boot Jack Prairie and listened to the varied thrushes claim the forest as their own—for MAXXAM, there are still trees out there, standing lumber.

"EPIC has filed for a temporary restraining order, but it hasn't been heard yet," Doug went on. "Do you have time to hike out there?"

It was just a few days before Christmas. I had knocked off work on the book and planned to simply relax, mail some packages, and wrap presents. Chuck was treating my children and me to dinner and Handel's *Messiah* in the evening. The last thing I wanted to do was go see more clearcuts sprayed with herbicides. But I had promised myself that I would see a tree cut down before this book is finished, so I said, "If I can get some packages mailed this morning . . . but I *have* to be back at five-thirty."

"No problem."

On the way up the dirt road that leads to Boot Jack Prairie, Doug stopped and peeled the backing off a bumper sticker that read, "Trees, America's Renewable Resource." He got out and stuck it to his back bumper and hopped back in the car.

"I like to get it dusty before I get there." This was his car's camouflage. "I've got a box of about five hundred of them someone gave me."

We parked the car on a rutted ranch road amid ancient lichen-draped oaks and dropped down onto Boot Jack Prairie. It felt strange to have come full circle and to be revisiting one of the groves, to feel at home there. We cut diagonally across the gigantic prairie, dropped into the grove, and stopped. Old-growth forest . . . it is like nowhere else on Earth. The equilibrium halts the soul, halts the feet midstep. How right for Christmas, for the day before the solstice, to come to this church and honor myself with the trees' presence. Hello, grandmothers . . . still standing here where I left you over a year ago . . . I am slightly wiser now.

Even though I knew that we would have to keep moving if I was to make it back for the *Messiah*, I gave myself a moment to relax in that damp cocoon of stability.

We made our way down to a logging road that cuts into the grove. Silt was suspended in the puddles as it had been a year ago. Now I know to look for this indicator of the recent passage of a vehicle. We dropped directly off the shoulder of the road again, continuing our descent of the slope like deer. We had entered residual forest, old-growth that had been thinned, the smaller trees left to prosper in the increased light.

"Stop . . . I hear saws!" Doug said. We held motionless. Yes, saws. We proceeded more cautiously, stopping to listen, even more like deer. But deer would have gone in the other direction. We continued on, toward our own kind.

We drew close, protected from view by tan oak and brush. We could smell the chain saw oil, the sawdust, and we could hear voices, though it was hard to discern words. We sat down in a safe place to survey the scene. Was the top of one of the trees directly in front of us shaking slightly? I thought so. Sometimes when I am out painting, I suddenly realize that a gopher, unaware of my presence, is working the roots of a nearby plant. I stop painting, fascinated by the subterranean demolition. Now, here, with some of the same unconsciousness but on a far larger scale, a whole tree was being worked on. I plotted the tree's possible trajectory, making sure we were safe, and then just sat, party to the killing.

"It doesn't seem right to just sit here," Doug said.

"Yeah," I said. I looked at the already disturbed landscape around us, a mess of tan oak and ceanothus competing to take advantage of the relatively recent flood of light. I looked at the upper trunk of the tree that I suspected was the quarry of the loggers out of sight below. It wasn't the massive tree that I had imagined would someday break my heart. It was actually rather small, a sort of anticlimax, I thought privately.

"That tree has stood here through storm after storm and now, in just a short time, it will fall," Doug said solemnly.

The whole landscape was such a far cry from an ancient forest that I began to wonder if it mattered.

"It's stood there," Doug went on, unaware of my ambivalence, "for four hundred years, five hundred years . . ."

"Just a baby," I thought.

"Since before the Pilgrims . . ." Doug went on.

"The Pilgrims?" I thought, and a light began to go on.

"Maybe since before the arrival of Columbus . . ."

"Columbus?" Sometimes I am so slow. In this ancient landscape, it has taken me so long to to grasp proportions. If a tree wasn't a seedling at the time of Christ, I think it is young.

"If they'd leave these trees alone, instead of cutting them all down and dousing the land with herbicide, in a few years they'd have something here again. These trees would be . . ."

The saws suddenly stopped, and along with the unseen loggers downslope, Doug and I paused in silence while the tree did what every other tree does that has had year after year of connective tissue severed. Its top began a smooth arc toward the ground. As soon as it landed with a final crash, the saws began on another tree. It was so fast! Because the trees had grown in the shade of the ancient forest, even though they were probably quite old, they were only a few feet in diameter. In fifty years their branches would have closed the canopy of the forest once more and the second growth beneath them could have been selectively harvested. In this way, with just a little restraint and planning, the forest could exist "into perpetuity." As it was, in just days all the trees around me would be gone.

Another tree began to shake almost imperceptibly.

I thought of the ad I saw recently in a local arts magazine, "Forestry, it's an Art. Pacific Lumber." This was not an art; it was simply a waste.

The second tree began to fall. I thought as it fell that it was about the size of the first tree, but it kept falling, making me realize that I was watching the top of a considerably taller tree than I had guessed. When it finally landed, I felt as if the Pilgrims had just crashed into Plymouth Rock and were mixed up with the wreckage of the tree. I felt as if the entire history of our country counted for nothing. I wanted to turn and run back to the grandmother trees and tell them, but they know. At exactly three-thirty the loggers packed up and left.

As soon as the sound of their trucks had faded into silence, we got up and made our way down through a chaos of trunks and branches. The cuts were sloppy, even dangerous. The goal of the cutting was clearly to get the trees down before the lawsuit commenced. We came to the freshly cut log of the tree we had seen felled. Beads of clear sap had risen on the outer few inches of the log. Further in I used the blade of my pocket knife as a marker to count the rings within one inch.

"Thirty-five years!" I said to Doug.

"Boy, that's a lot."

"It must have been growing in dense shade for the rings to be so close together." We multiplied and came up with 840 years. I double-checked my count and had no doubt. Eight hundred and forty years. In the year 1000, feudalism, with its knights and castles, dominated Europe. In 1040 the development of a magnetic needle opened the way to the invention of the compass and to the eventual exploration of the New World. When this tree was a seedling, men were just instituting the concept of banking.

Doug and I stood in silence. He laid his hand on the log, silently making amends.

I looked at my watch.

"I have to get back, Doug. I have to be home by six o'clock at the latest."

"We can walk back on the road. That'll be faster and you can see the latest herbicide spraying."

No sooner had we started up the road, however, than we saw a person scurry across in front of us. We dove into the bushes. We heard hoots from up on the hill and realized that the person was an Earth First! activist, not a logger. We kept walking. Doug heard voices. We dove into the bushes again. Doug crept low to peek over the hill, and four pickup trucks were parked in the road ahead.

"That's Carl Anderson," Doug said. Carl Anderson is the head of PL security; it was he who had put the butcher paper over the windows during the pepper spraying. He was also a candidate for sheriff in the last election. We later learned that an activist named Geronimo had been captured and the others were trying to distract the men so Geronimo could be released from the pickup truck.

We quickly turned around and retraced our steps back up the tangled slope until, in the semidarkness, we once again stood among the ancient trees of Owl Creek Grove. Holy trees. O holy night . . .

"I'm not going to make it to the *Messiah*, am I?" I asked Doug quietly.

"I'm afraid not."

The backtracking had cost us too much time. Doug had been counting on being able to take the road.

Doug leaned forward and put his face against a huge tree, but the nearest tree to me was just a foot in diameter. Doug's action made me want to put out my own hand and touch the flank of one of the giants, but there was still some hope of making the *Messiah*, so I didn't want to backtrack. Instead, I studied my little tree the way someone might regard a disappointing blind date, and then I asked myself, "What can I learn from this tree?" It was so insignificant, virtually unnoticeable in a place where size and age are everything. And then it suddenly reminded me of one of my children standing there, and a shift occurred inside me.

"This forest isn't just for the big trees. It's for the little trees as well."

While I had thought about the forests' future, I hadn't felt it. I was looking at a tree that had the genetic capability to live a thousand, two thousand years, into a time we can't even imagine. How cavalier we are with the future, with future lives we do not own. We can hoard the present but we have no feeling for the beings whose ancestors we are.

The forest had an extra dimension. I was reminded of a time when I was a child looking up at the stars out in the desert. Instead of seeing the sky as a flat ceiling, I suddenly saw the stars spatially, the little stars distant and the bigger stars nearer to me. Though this is not strictly accurate, the sky gained a new dimension of depth. This is what happened to me in the forest. Instead of seeing it flatly, stereotypically, linearly, imbued only with the resonance of the past, I suddenly saw time spatially, the past, the future, standing all around me, embracing me in the moment.

I didn't make it to the *Messiah*. Chuck took my children. It had been my idea, but they came back radiant, excited. The performance had touched them, engulfed them. I felt sad to have taken this big bite out of my own Christmas. "Something's lost but something's gained in living every day." These words from the Joni Mitchell song come back to me more and more. I had paused momentarily in a silent, ancient forest at dusk before scaling Boot Jack Prairie as fast as I could, still hopeful of making the concert. Later, when we all compared notes, my daughter stood before me still dressed up, nearly a woman. I had made a sacrifice, and the silence of time was my Christmas.

Little brown bats

CHAPTER 75 RESULTS

LAST night I came back from a weekend trip out of town and found that the headlines of the *Times-Standard* read, "State Says PL License in Jeopardy." According to the article, Pacific Lumber may have trouble renewing its license to harvest timber as of January 1, 1998. After a weekend spent with my heart so heavy I could hardly sleep at night because of the incessant rain and because this struggle seems futile, this change is stunning news. I think of the words of an Indian philosopher, Rabindranath Tagore, that I memorized when I was in high school and which I often paraphrase in my mind like a mantra: "Nothing is lost to eternity, everything blooms at its hour; Sick at heart I fell asleep and woke to find my garden full of flowers." Often we think that our work is in vain, but really its time has not yet come. We work for the forest, for the Earth, for God, and it does not diminish that work if no one ever knows what we contributed. That work fills our soul and creates our own sense of integrity. The trees give us our very breath, and they do not ask for credit. And someday, perhaps even after we are dead, that energy blooms beyond our knowing as a gift for others, two-legged, four-legged, six-legged, hundred-legged . . . as a gift for the millipedes that sometimes bloom in the redwood forest, and we are once again reborn.

The article quotes Gerald Ahlstrom, deputy chief of the CDF Forest Practice Enforcement and Litigation Division, "'There are about two thousand licensed timber operators in California. We've had to send these notifications to just eleven of them. It's something we only do as a next-to-last resort, when there are repeated, deliberate violations, and there seems to be no other way to obtain compliance.' Ahlstrom said about half of those notified actually lose their licenses. Most are small loggers.

"While he knows of no precedent for revoking the license of a major timber company, Ahlstrom said it's not unthinkable. 'Law enforcement has to be evenhanded,' he said. 'Big companies can't be given immunity.' "

In July of 1997, PL was put on probation. As a condition of its probation, it was obligated to commit no further violations. Yet "between August 1 and November 10 CDF inspectors cited the company for a total of 31 additional violations, some of that were deemed uncorrectable." Of course, anyone who has walked PL land knows that for every citation there are hundreds of unnoticed violations. Among those that are "uncorrectable" the article mentions "clear-cutting an area that was supposed to be thinned and improper road-building that caused a slide and dumped soil into a watercourse." From the viewpoint of the animal residents of the forest, this may translate to "cutting down old-growth trees that might have provided precious nest limbs for marbled murrelets for centuries to come, or clogging a stream with silt that might prevent the last stragglers of a once abundant stock of salmon from ever spawning again."

During this past year I have watched Kristi and Ralph and Mike lift their heads, look around and say, "This is not right," and act. They went to the Board of Forestry and asked for an emergency ban on winter logging. Though the Board of Forestry did not agree to that emergency ban, now, just a few weeks later, we may have something better and our "garden" may be suddenly "full of flowers."

I am not so naive as to think that this victory, if and when it comes, will be a lasting one. Erosion of all kinds appears to be constant. But that victory may make another victory more imaginable. And victories here, in Huboldt County, may give courage to those who work for balance elsewhere. Just a month or two ago, it would have been inconceivable for the *Times-Standard* to be writing articles that so clearly define the issue. Now letters to the editor cover page after page as the people themselves become empowered to speak and act.

Red alder

CHAPTER 76 METAMORPHOSIS

RECENTLY a biology professor lent me several bird mounts for use in my drawing classes. I took a male wood duck to class, and the children sat around the duck and drew it from different angles. As one might imagine, each child had a different experience, and no two drawings were the same. I am quick to point out to students when they all draw the same subject that their drawings are, first of all, portraits of themselves born out of a unique moment in their lives. If they drew the same wood duck again tomorrow, their new drawings would be portraits of a new day, and of new children. The drawings are records of the metamorphosis of the children's awareness.

So it is with this book. I am surprised to realize that it is a story of my own metamorphosis. I look back on my life before this began the way a frog might look back on the experience of being a tadpole. My mind rationally connects, but the fact is that ever since the mountainside of Shaw Creek Grove reached out and drew me to its damp, cold side, I have become someone I like better and understand more. I look at cold from a less self-centered point of view. I feel the rain and fog with deeper reverence. I clench my human teeth tighter and recognize the value of fighting for what is right.

Someone said to me, "It isn't just Hurwitz. It's not fair to criticize him." And I thought, it *is* Hurwitz, and it *is* fair to criticize him. But we must see the Charles Hurwitz within each of us and constructively criticize ourselves as well. The term "cumulative effects" has taken on personal meaning for me. Our whole earth is suffering from the cumulative effects of a million minute daily actions.

The word yoga means "wholeness." Western religion and Western medicine encourage us to believe that we will decline toward sin and ill health without constant vigilance—entropy. Yoga's basic premise is that our natural inclination is to return to a state of wholeness and health—Gaia. I feel that I have been put on a path toward greater wholeness, not by my own doing necessarily, but by the influence of the forest itself, and the people I have met. Their wholeness has given me an increasing distaste for inconsistencies in my own life that are the result of "splits" that I have voluntarily or ignorantly accepted. Gradually those splits have been merging, as if I were looking through binoculars that are pulling two images into one.

I would be limiting your imagination if I told you what you "should do." All I can say is that we must all brainstorm, alone and with others, opening our minds to new visions of our personal lives and our life as a nation. After my interview with Spring, she buckled on her bicycle helmet and peddled away. She does not own a car. Her father doesn't either. Together with friends they tore up their driveway and planted a vegetable garden. "Find a corner of the world and fix it."

But we must also look at the big picture—nationally, globally. We must take responsibility for more than just our personal sphere. I kept asking Ken Miller what we were asking the members of Taxpayers for Headwaters to *do*.

"I don't *know*," he kept replying. "*They* have to think of it."

"But we have to *tell* them," I countered.

"No we *don't*," he argued back. "It is better if *they* think of new ideas."

We argued for weeks, and then I suddenly realized he was right. If citizens truly think they can make a difference and sincerely brainstorm, we can awaken the intelligence of the nation and not just depend on a few leaders. Whether it is tearing up your driveway or focusing on a local issue of concern, it is best if your own open mind is your inspiration and guide. The key is that we feel empowered not just to act but to think, critically and with hope. We have no choice. Otherwise we will find our children forced to dangle on cracking doors high in the redwoods to draw attention to problems we are neglecting.

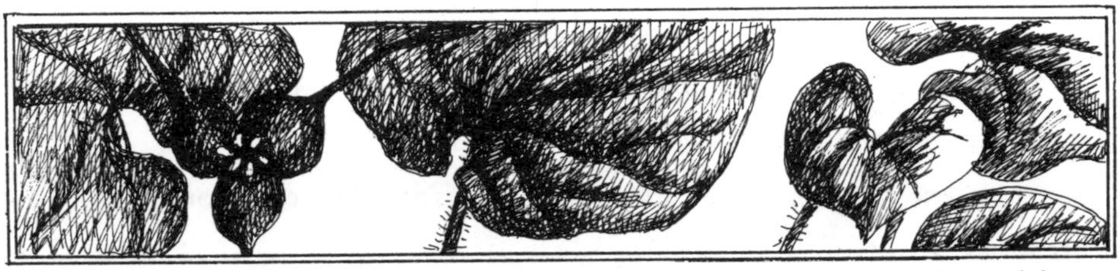

Long-tailed ginger

CHAPTER 77 KNOTS

WHEN I walked by the television the other night, I caught sight of an ad that talked about greatness. I paused in the semi-darkness behind my children who were curled up on the couch. It described people who have "lived as if they could change the world . . . and have."

As I write this a young woman named Butterfly has just celebrated six months' occupation of an ancient giant in the Timber Harvest Plan adjacent to the Stafford slide. Each day from her treetop she communicates by cell phone to reporters all around the world, using her visibility to speak about sustainable forestry and the ecological crimes being committed around her. She answers stacks of mail that are delivered along with food and supplies by her ground support, other members of Earth First! She writes back to classrooms full of children who have written to her thanking her for trying to save the forests of their planet. For thousands, Butterfly is becoming a mythic heroine.

I decided to try to interview Butterfly by cell phone. One rainy afternoon I was about to help David with his paper route when I said, "Let's go get a mocha first. It'll just take a minute."

As I waited for our mochas at a local coffee shop, I noticed that some people were just ending a meeting by holding hands.

"That's civilized," I thought to myself, before I actually looked at their faces and realized I had happened upon an Earth First! meeting. Spring was there and a fellow named Josh Brown. After Josh introduced me to the group and told them about my book, I said, "I'd like to interview Butterfly by cell phone."

"No problem," Josh said. "Do you have a piece of paper? I'll give you the number."

"Why don't you go up there?" Spring interjected lightly.

"You mean go up the tree?" I asked.

"Sure," she said with pixie-like ease, as if I might fly there on transparent wings.

"Oh, no, I hate heights. And I don't climb trees," I said, matter-of-factly categorizing myself. I feel pretty sure that a fear of heights is genetically transmitted. If it isn't, then my fear of heights stems from all of the times I have seen my mother recoil from the railings of balconies and the rims of canyons. I have always prided myself on the solidness of my resolve to remain in firm contact with the ground.

Next to me, like a peaceful guardian angel, there appeared a mild-mannered young man with an Abe Lincoln-style beard.

"I can show you some knots if you like." His voice was so calm that it seemed to still the air in the room rather than add to its invisible currents.

"Knots?" I asked. What did knots have to do with climbing? I am always looking for male influence for David, however, so I said, "Uh, when?"

"Right now," he said. "We can sit at a table. I have my ropes in the truck."

David and I conferred, estimating that we could spare some time and still get his papers delivered.

"My name's Sawyer. I'll get my ropes."

Sawyer proved to be an extremely systematic and thorough teacher. Furthermore, tying knots was fun. During the nearly two hours that we sat at our table wrapping the nylon-reinforced loops of rope called prusicks and knotting them so they would slide smoothly up and down the climbing rope; memorizing the order of our caribiners on the harness and double-checking to make sure they were tightened but not too tight; slip-knotting the loop of webbing through which we would slip our imaginary feet; and reviewing over and over how to keep our equipment in proper order to reduce the chance of entanglement mid-climb, I felt a little guilty that I would not actually be climbing the tree. Sawyer was so sincere that I felt a little devious.

"Like this?" David asked, slipping his well-tied knot along his length of rope.

"Perfect," Sawyer praised.

As the lesson went on, however, I found myself growing eager for another. Knot-tying is something a visual person can excel at. The whole system seemed like a reasonable way to perhaps conquer my fear of heights. As we packed up to leave the cafe, Sawyer said quietly, "Next we can go climb trees in the Community Forest."

"Uh. . . ." I had thought the next lesson would involve advanced knot-tying. This was it? I was ready to climb? Feeling sure that I could eventually back out of the experience with countless legitimate excuses I had already inventoried while I practiced the knots, I said casually, "I have some redwoods in my backyard. We could use those."

"Great!" Sawyer said and a morning was agreed upon.

Sawyer arrived promptly, and we went out to the trees in a drizzling rain.

"Nice," he said. "These are perfect." He found a little piece of firewood, secured his climbing rope to it with a thinner cord, and skillfully threw it over the lowest branch, twenty feet up. In moments he was up in the tree securing the rope and down again, giving me a refresher lesson on my equipment and then actually demonstrating how it all worked. He slipped his legs into a harness, tightened the belt, attached his lines with carbiners, double-checked all his equipment, then quickly ascended the rope and rappelled back down to the ground.

I eyed the lowest branch. Either I had to make it to that branch where I could stand and switch over to my metal figure-eight so I could rappel down, or I would have to switch over mid-rope. This latter maneuver involved a rather sickening moment when I would have to cross the rope over my body and yank on it so that all of my weight would suddenly be held in the grip of my own hand with the rope pulled tight against my thigh. This was not as dangerous as it seemed, since I would always have a back-up safety line attached to the rope. But it looked dangerous.

I harnessed up and began climbing. The first thing I noticed was that my enjoyment of the system was sending me very quickly up the rope. I merely had to raise myself to a standing position in my foot loop, to give me slack on my harness line so that I could then slide its knot upward. Then I would sit comfortably in the harness while I slid the knot of my foot loop upwards. By repeating this two-step process, I ascended like a well-practiced inch worm. But, because of my human scale, I rose not by inches but several feet at a time.

"Good!" Sawyer called up from the ground.

I looked down. He was far below! I looked up. I was nearly to the first branch! I felt the possibility of panic. What if I really did panic and couldn't get down? I ruled out panic. I decided it would be wise, before I got any higher, to try rappelling from mid-rope. I might not be able to get onto a branch in order to rappel.

"I'm going to try switching over to the eight and rappelling down."

"Good!"

But first I dangled and looked around the neighborhood over the rooftops. I forced myself to relax. Indeed, even without consciously holding myself up, I did not fall. I checked my knots. They all looked "clean" and functional. I forced myself to look down into my own yard.

"Just sit here until you can come to peace with this," I thought. I was changing my image of myself. I just might be someone who left the ground after all.

"Try swinging," Sawyer suggested.

"Swinging?"

"Yeah. It's fun. Push off from the tree and just swing. It'll give you confidence in the rope. It's securely tied around that very strong branch and I wrapped the rope around the tree as well, so there's no chance of it slipping. You're completely safe."

So I swung. I was not only a climber, I could play in the air. I looked over at my neighbors' second-story bedroom window.

"I'm a climber," I informed them silently in case they were watching.

I came to rest again with my foot on the trunk of my tree and looked at its fibrous, red bark with new affection. Hello tree. I looked up into its canopy and suddenly it appeared to be a whole world waiting to be explored. Hello, canopy. My yard has gained a new dimension—up.

I switched over onto my eight with relatively little difficulty, and began to let out the line through my right hand which was held down against my thigh. Miraculously I began to slide smoothly down the rope. I clinched my fist and stopped. I opened it slightly and slid. I clinched and stopped. I had become a climber who could get herself back to the ground, and did, in just seconds.

CHAPTER 78 MYTHIC REALITY

JUST after the mud slide destroyed the community of Stafford and a new Timber Harvest Plan had been promptly approved for the adjacent, equally steep slope, a young man named Daniel took it upon himself to climb the mountain and investigate. As he explored, he found an enormous ancient redwood standing on the ridge. Its top had broken off—who knows when — during a thunderstorm a hundred, five hundred years ago? In the way that redwoods have of making themselves malleable to the demands of the centuries, a new top grew, with a natural platform at its base. Smaller redwoods grew upward from the giant's roots. Daniel, bare-footed, free-climbed one of these smaller trees until he could cross over into the huge tree's branches. Then he ascended, free-climbing without a rope until he was 180 feet off the ground.

There he took up residence in a cave in the trunk of the tree, picking and eating huckleberries and salmon berries from the shrubs that grow in the soil lodged in the crotches and crevices of the tree's gigantic branches. He erected a beacon light that was visible to the residents in Stafford and motorists on Highway 101. Members of Earth First! climbed the mountain to find the light's source, met Daniel, and returned on the full moon, christening the tree "Luna," and making plans to erect a platform high in her branches. Heavy building materials were lugged up through the forest to the ridge top, then pulled by a rope into the tree's canopy. An eight-foot-by-eight-foot platform with a roof of blue tarps was completed, and the Earth First!ers began a round-the-clock tree-sit to prevent Luna from being cut down and to protest the clear cutting of MAXXAM's forests, the logging of old-growth trees, and the destruction of Stafford.

While other people took their watch in the tree, Daniel went hiking on The Lost Coast, twenty-six miles of roadless coastline south of Eureka, where he met Julia Hill, the twenty-three-year-old daughter of an Arkansas itinerant minister. Julia was recovering from severe injuries she had suffered when a drunk driver struck her car from behind, jamming the steering wheel through her eye socket into her brain. She had spent two years relearning how to form words into sentences. When she met Daniel, she was attempting to reconcile the new psychic awareness that had developed after the accident with the teachings of her childhood.

Daniel told Julia about the ancient redwoods, and about the tree near Stafford. She hiked up the ridge to see Luna, learned to rope-climb, and went to Luna's top.

"I went up into Luna intending to stay for two weeks," Julia tells reporters, "and I never went down." She took the "woods name" of Butterfly. Now it is Butterfly who turns on the blinking beacon light each evening for all the world to see. And, increasingly, the world is looking. It is asking, "Why does a beautiful, intelligent, well-spoken girl halt her life and begin it anew 180 feet up in a tree? And how long will she stay?"

CHAPTER 79 BUTTERFLY

SAWYER asked me if I was ready to go up Luna. Twice I stalled, saying I didn't think I was ready. Finally he drew me a large picture of the tree with the platform near the top, and as a finishing touch he put in a little woman on a rope.

"That's you, rappeling down."

"This is what Luna looks like?" I asked.

"Yeah."

I eyed the first branches and the scale of the little climber. "*Just* like this?"

"Well, the scale's a little off." He studied his own drawing. He had even put in a little compass design showing north and south. "That's sort of nice, don't you think?" he asked hesitantly, pointing to the little arrow and the compass points shown in three-dimensional relief. "I used to draw a lot. I'd like to get back into it."

He gave me the drawing, and I scotch-taped it to my kitchen cabinet. I gave him the pad of charcoal paper, and he thanked me with reverent appreciation. I studied the drawing as I made dinners and did dishes. Luna was gnarled and friendly looking. I looked competent. Finally I said to Sawyer, "Okay."

"Friday?"

Doug said he was coming and did some hasty practicing with Sawyer. He didn't seem too worried, although he'd never done any rope climbing before.

Friday came, and we stopped at Earth First!'s Media House and picked up Sawyer, Shakespeare, Molly, and a reporter named Chris, from a Berkeley newspaper. I find I am no longer surprised when I am introduced to a group of people, and many of them have self-chosen "woods names" like Garlic, Slo-mo, Spruce, Tofu, Treetop, Daves-not-here, Felony. Felony took some time. Melanie . . . Felony . . . Melanie . . . I played with the sounds in my mind. Most of the names seemed merely playful, even childish. As I have gotten to know the people behind the names, however, a shift has occurred for me. In real life I have found Felony to be a precise, cheerful woman who reminds me of a hardy Girl Scout leader. And as I have heard young people stoically refer to court dates, jail time, large fines that they have incurred as the result of nonviolent action in defense of the environment, her name has begun to fit the mythic quality of this struggle. All of the names, whether outrageous or innocent, have opened my mind a little more and have made me begin to look quizzically at people still named Bill or Bob or Nancy. Are we doing enough?

We hiked across the freeway and then climbed for two hours to reach the ridge. Four times we had to throw ourselves into the bushes to avoid logging and security trucks until we finally hit skid trails that took us straight up the mountain. In order to keep pace with five young people in their twenties I kept my face down and simply breathed. Remnants of forest understory, which had regrown after the passage of the bulldozers, passed beneath my gaze—yellow redwood violets, an occasional wild bleeding heart, wood sorrel, sword fern, salal, and, protected in the lee of giant stumps where the bulldozers hadn't scraped, wild ginger with its rounded, heart-shaped, fragrant leaves.

At noon, Sawyer paused on a landing.

"Anyone ready for lunch?"

"Sure."

"I need a cutting board," he said looking around, and chose a barkless log angled up a gravel slope. He and Shakespeare and Molly took out the loose vegetables they had cradled, bagless, in their arms, from the grocery store to my car. Now they bellied up to the log and began slicing, moving up the log so that a buffet of hazelnut bread, clusters of avocado, tomatoes, a neat pile of spinach leaves, minced raw garlic, sprouts, and sliced onions was laid out by the time Sawyer unwrapped some slices of Swiss cheese and stood back.

"Help yourselves to sandwiches."

I was charmed by the simplicity of this meal, laid out on the flank of one of the vanquished. Such order in the midst of waste.

As I ate, I could hear the reporter from Berkeley talking into his recorder, ". . . barren gravel slopes . . . sixty-year-old second growth . . . What are these trees?"

I looked up. "Grand fir," I said.

"... grand fir ... Stafford slide ..."

Indeed, we could look over just eye-level with the top of the slide that buried the community of Stafford. I had never thought that I would actually see the origin of that slide.

"We'd better keep moving. Joan and Doug still have their climb to do and it's getting late." Sawyer had wanted to leave at dawn, but Doug and I were late. Sawyer still refers to himself in present tense as an Eagle Scout.

"You're an Eagle Scout?" I asked. "You use the present tense for that?"

"Yes, once you're an Eagle Scout, you're always one. Scouting is how I know how to do all the things I do for Earth First!"

What irony. When I was a child we spent hot summers camped on the banks of the Colorado River. One day my father informed my brother and me contemptuously that seven Boy Scouts had gone over Squaw Dam and died. My father, ever an anti-establishment kind of guy, used this as evidence for the rest of my childhood that scouting was a waste of time.

"That's how Boy Scouts are," he would say. "They're taught how to camp, but not taught common sense."

Had I been older, I might have countered that the quality of any scouting experience probably depends on the commitment and experience of the leader and the enthusiasm of the scouts. Now, after watching Sawyer in action, I have found myself listening to other parents' conversations for mention of a potential troop for David.

Sawyer bagged up the vegetable scraps and put them in his pack. Leave no trace, I thought, as if we were in a wilderness. Then we headed straight uphill again for another half-hour climb. Finally Sawyer stopped and said solemnly, "There's Luna."

Now that she is isolated by the surrounding clearcut, the tree has been described as "a fist" raised high on the ridge. I saw her as a beautiful, unique, complex vertical landscape. I could barely see the platform where Julia was sitting and where she is still sitting as I write this.

"Hello, Julia ... Butterfly. Hello, Luna," I said silently to the distant tree.

I bent my head again to an even steeper climb. Talk was impossible. I am nearly fifty. A year and a half of nearly non-stop writing have taken their toll.

"I'm going to see if we're within range," Sawyer said, pausing to try the two-way radio. "Julia, are you there? Hello, Julia. We're on our way up."

"Yes! Hello!," (delighted laughter). "Hello!"

As we had hiked, I had gotten to know Molly, a solid, sweet young girl in her late teens or early twenties. She was walking just ahead of me. As I crested the last rise and started down to Luna, I almost tripped over Molly, collapsed on a stump. The rest of us had either never seen the forest that had once surrounded Luna, or else had already revisited the area after the helicopter logging. Only Molly was being hit with the brutal reality of before and after. She had her face down, and was unmoving, except for her sobbing. I laid a hand on her shoulder.

"I'm sorry," I said.

But I had to keep moving. It was getting late.

"Joan, are you going up first?" Sawyer asked.

"Sure," I said. Why not? What does it matter when you do something absolutely insane?

So I put on the harness, tied my knots, hooked on my caribiners exactly as I had in my back yard—only this rope was 180 feet long. The first branch was thirty or forty feet up. The rope disappeared beyond it. The top, the platform, Butterfly were nowhere in sight.

"Inch by inch," I said to myself and started climbing.

"Doin' great . . ." was the last I heard from Sawyer before I entered that silent no man's land that must be a familiar reality for climbers. It was just me, the silent spider methodically ascending her web. I looked at Luna's trunk. She was there like my only friend. What had she seen in her thousand-plus years? Most likely the miracle of a steady continuum, the miracle of sameness, non-violence. She had simply stood in the company of her kind, visited most days only by fog.

What had it been like to have the saws suddenly arrive and begin felling the trees around her, and gigantic two-prop helicopters arrive, tearing the mats of vegetation from her branches with their wind and lifting the other elders out in truncated logs? What had it been like to suddenly be exposed to El Niño's ninety-mile-per-hour winds, roaring in gusts across the barren slopes of the mountainside after more than a century of companionship with her kind? Now Luna stands isolated by the clear-cutting, with a view of the Eel River Valley that I doubt she ever had before. It was easy to personify Luna, hanging there by her side. I looked north at MAXXAM's mill in Scotia, its log decks coming into view as I climbed.

"We are on the countdown to eternity," a radio announcer's peppy voice came into my mind from the day before when I caught a moment of a religious broadcast on my car radio as it scanned the local stations. Eternity had produced Luna, and the embracing of annihilation by our culture had transformed Butterfly into an inspired spokesperson for the tree. Both the tree's and Butterfly's existences seemed equally inscrutable to me. I felt like I was going to visit the wise woman of the mountain . . . stand, slip the knot, sit, slip the knot . . . stand, slip the knot, sit . . . don't look down . . . stand, slip the knot . . .

I kept climbing far longer than I'd thought I would. How conservative was the scale of Sawyer's drawing? Apparently I hadn't given him a long enough piece of paper. Where was the platform? Don't look down. Don't use your imagination . . . stay in the present moment. I knew I had long ago lost sight of the ground, but I couldn't look . . .

"Hello! Joan! You're almost here!" a voice suddenly called cheerfully from above. I couldn't see Butterfly or the platform, but apparently she had seen me.

"Almost there," I coached myself. "Don't look down. It'll all be okay when you get on the platform."

But when I got up to Butterfly she said, "Hi, Joan. I'll give you a hug when we get to the top. But we have a little bit of climbing to do first."

"What?"

"I'm going to unclip you from your rope. You have your safety on. You're perfectly safe. And I am going to clip you on with these two lobster claws and move them with you as I show you where to put your feet and hands. You will be attached to the tree at two points at all times. You're perfectly safe . . ."

"I have to climb through the branches? I asked incredulously. "No one told me about this."

"Just a little ways. Now I'll move this lobster claw forward and clip it in this loop and you reach and take hold here, then move your right foot out and place it here on this branch . . ."

I was overwhelmed. I was stepping across mid-air from branch to branch. Butterfly was not connected to a rope at all. She was bare-footed, ushering me step by step through the tree. On one particularly long stretch I said, "I can't do this."

"Well, ma'am, you have to."

Ma'am? She was right. I had no choice.

Then came a rope ladder! I was going up that?

"Now just put your hand on this rung..."

"No one told me about a rope ladder! I can't go up that rope ladder. I'm really phobic about heights!"

"Well, ma'am, you just have to." She had a point. I did want to get to the platform which was still about ten feet above me. I was getting cold. In fact, I was starting to shiver.

"Put your right hand on this rung. You're going to go up two rungs and then step across onto this branch. You're clipped on. There's no way you can fall . . ."

At last I made it to the platform and hoisted myself onto it. I received the promised hug and looked into the beautiful, naturally jubilant face of Butterfly.

She communicated on the radio with Sawyer.

"Joan's here. Is Doug on his way up?"

"Yes. He should be up there in about twenty minutes. We're tying his pack of photographic equipment to the supply line now."

"Okay, tell me when you're ready for me to start pulling it up."

Just then another voice could be heard on Butterfly's radio. Butterfly said, "Excuse me for a moment. There's been this logger who's been cutting in on my signal."

I sat on her sleeping bag shivering, looking around at her belongings hung in little bags and tucked between branches in precise order. There was a little one-burner propane stove, a lighter, a pot, a knife, a dish scrubber, a candle held in a little reflective holder backed with aluminum foil, some books . . . I made the mistake of lifting the blue tarp slightly and peeking over the edge of the plywood on which I sat. Oh. There was the Eel River Valley. I had thought the platform would have been a little more than just two pieces of plywood and some blue tarps. I expected something more substantial, befitting a shrine. But this was a young shrine, just fresh from being merely a treetop. The blue tarps were held up by Luna's own branches and twigs whose tips were wrapped with duct tape to keep them from shredding the tarps. The tarps were lashed down with heavy ropes. I listened to the wind. It was picking up. It was about four o'clock in the afternoon. I wished Doug would arrive so someone else would be scared besides me, and so that we could leave. Butterfly meanwhile explained to the logger that she really was in the top of a 180 foot redwood tree. I could hear the man's wife in the background saying, "Honey, that's that woman whose been up in that tree . . ."

Butterfly reached over and pulled out a stack of cracker and cereal box fronts and backs which she had blanket-stitched around the edges with yarn.

'This is some of my poetry and drawings. They bring me these boxes to write on. Here, are you cold? Do you want my blanket."

I curled up in her rough army blanket with the box lids while she dangled her bare feet off the platform and continued debating with the logger on the two-way radio.

"We're not against logging," she patiently rebutted. "We're not against logging at all. MAXXAM is making it look like they're providing all of these jobs, but, unfortunately in another few years there's not going to be anything left." The nice thing about a debate over a two-way radio is that only one person can talk at a time. Technology provides a built-in formality.

"Clear-cutting is the only way you're going to be able to get all of those old trees cleared out of there and begin to plant the trees that are actually going to grow," the logger said.

"Once again, you're only talking about the trees. From here I can see the tree farms that are the answer to these clearcuts, and I guess they're going to harvest them on a sixty- to seventy-year rotation and the soil is going to be massively depleted just like it is on farms where they plant the same crop year after year and the soil gets so depleted they have to start dumping chemicals into the ground to try to supply what would naturally be in the soil."

I was amazed at how much studying Julia had done.

"If you have a chance, if you haven't read it yet," she went on, "try to get your hands on the Headwaters Stewardship Plan. It is a plan for how this forest can be selectively logged . . . sustainably. We want to work with loggers. In fact a lot of loggers have had a lot of input on the Headwaters Stewardship Plan."

"Yeah, well I'd like to take my chain saw out there and select a few big ones."

Meanwhile I was reading Butterfly's poetry. . . . "It is a desperate picture these branches frame . . . I want to strike out at the ones to blame . . . but that won't heal this sadness too deep to name. . . ." In the margins were images of goddesses whose bodies grew out of the trunks of trees. I recognized similarities to pictures that I have drawn in the past, yet Butterfly has taken the power of raw idealism into direct action in the real world to challenge the behavior of a major corporation. At an age when I was in college studying Baudelaire, or T. S. Eliot, Monet or Van Gogh, she debates the merits of sustainable forestry from high on her ridge top and grants interviews to the *New York Times* and CNN.

Finally the debate with the logger ended. I admired the way she thoroughly communicates with each person who enters her sphere, whether by walkie-talkie, letter, cell phone, or rope. I studied her physical beauty as we talked. She is dark-haired, clear-eyed, feminine yet boyish.

"I grew up with two brothers. I had to be tough," she said. "My father and mother taught me to speak out for what I believe in, but I think they worry they overdid it."

She described efforts by MAXXAM to get her out of the tree.

"They tried to starve me out with a round-the-clock, ten-day blockade of my ground support but I made friends with one of the security guards who was camped at the base of the tree. Most of the time I've been here I've just had obscenities yelled at me, but this is a person who has actually communicated with me and listened to me and talked with me, many, many times."

She described the helicopter logging, "I was led to believe that helicopter logging was a better way to do things. After being up here, I don't think it is. The difference is that the bulldozing for roads destroys the ground, and the helicopter logging destroys the tops of the trees. The helicopter creates intense down-drafts that take the trees and twist them like a tornado. The whole top of the canopy is shredded and blown into the air."

Indeed, one of the reasons I have always wanted to explore the canopy of an ancient tree is to see the mats of vegetation that grow there and have their own unique fauna. I had been disappointed that Luna's branches were stripped.

Butterfly told how, even though there was a prescribed safety zone around the tree during the helicopter logging of its neighbors, the blade of one of the giant choppers was brought in right at eye-level to the platform.

"I thought they were going to crash they came in so close. I know they were trying to scare me out."

Butterfly began to describe what it was like to endure the ninety-mile-per-hour winds that we experienced during several violent winter storms. Then she stopped to haul Doug's pack over the edge of the platform. And then word came on the radio that Doug himself was probably nearing the top of the rope.

"Excuse me, I'll just go help him up."

Climbing the tree was not exactly my idea, it was Spring's, but when Doug finally arrived and confessed repeatedly in front of this beautiful and eligible young woman that he was more terrified than he had ever been in his life, I was actually kind of pleased to learn that he blamed me.

"You got me into to this!" he said, laughing nervously and eyeing the platform around him as if he still weren't sure he had arrived. "But I guess you're getting me back for all the adventures I've put you through."

"Yeah," I said with satisfaction. I'd take the point in my favor.

"I'm freezing," Doug said. "Is this wind going to be a problem going down?"

"Wind?" Butterfly asked. "Oh, I guess there's a little wind. No, it won't be a problem. You'll be down in minutes when you go."

I had never seen Doug scared. I admired how openly he admitted to his terror. Since our climb up Luna, he has gone up equally high trees in Bell Creek and said this was easy compared to Luna, because they weren't isolated on a high ridge. It had taken us each about half an hour to ascend the rope. I actually fully expected that I would never get down out of the tree. As I listened to the wind, I was eyeing the eight-foot-square platform figuring out where I would sleep and how I would stay warm if Butterfly suggested we should stay until morning. On the other hand, I didn't want to be in the tree if a storm picked up. A friend who later spent a couple of days in Luna spoke of feeling seasick from the rocking.

Butterfly continued describing what it was like to experience ninety-mile-an-hour winds, alone, several nights in a row.

"The first night I was so scared I thought I was going to die. Branches were falling all around me. Suddenly the entire platform lifted up and threw me against Luna's trunk. I lay there hanging onto her and crying, 'Luna, I'm sorry to bring these bad feelings up into your branches, but I think I'm going to die unless you can help me get through this.' Then she said to me, 'If you're rigid like a sky scraper, you won't make it. Look around you at the other trees. They move and howl with the wind.' So I loosened up my body and howled at the top of my lungs and when morning came, I had made it."

I looked over at Doug's face. He was fascinated, while she wove together stories of her childhood as the daughter of a minister; described the horror of the automobile accident that cut short her plans to go to law school, and then told us about her first visit to the redwoods.

"The cathedral of the redwoods brought me to my knees like no man behind a pulpit ever could."

She told about the end of the ten-day blockade of supplies and how twenty Earth First!ers finally stormed the tree and sent supplies up her rope.

"It was so beautiful," she said. "They were so brave." She talked about being harassed by air horns, barking dogs, floodlights, and whistles to keep her from sleeping. But Butterfly has endured. I wondered how long Luna would endure.

"... even in the end, if I can't save her physically, I want to be able to save the memory of what she stands for. And what she stands for is our fight against corporate greed and the destruction it is causing our mother, Earth."

Doug said, "I'd like to get some photographs of you."

Butterfly went out into Luna's branches and looked back at the camera, smiling. I too was fascinated. Her beauty seemed to have so many aspects. One minute she was a tomboy, the next a gorgeous woman. Beyond her were mountains with clearcuts and in the distance the snow-laden Trinity Alps. The view was unbelievable. As she moved around the treetop the view behind her changed. It was a photographer's dream shoot. Then she came inside and Doug took some close-up pictures of her face. We talked a little more and then Butterfly said, "I think you should go. You're losing your light." Easier said than done, I thought.

"Who's going first?" she asked. Doug and I looked at each other.

"I will," I said. It was like agreeing to go first to our execution. I slid to the edge of the thin plywood, and Butterfly reversed the process she had gone through to get me up through the branches. I have always been better at going up than going down. In order to see where she was indicating that I should put my foot, I had to look down. If a view had the power to kill, I'd be dead. I had enjoyed looking out at the horizon from the tree, but not down into the distance through which I might actually fall. But the branch held, as did the next and the next.

When I got back to my climbing rope, Butterfly unclipped me from the lobster claws and back onto the rope, saying, "Now, put your feet on these branches on either side of the rope, squat down and let your weight go onto the eight."

I started to do what she said.

"I can't!" I said, standing back upright.

"I know. This is the tough part, but ma'am, you just have to."

Something about the "ma'am" made me sense she wasn't going to baby me through this either. I tried again.

"I can't!" I said again, straightening up.

"I know, ma'am, it seems like you won't be able to hold your weight with the rope but you will. Everybody thinks this. Just sit down and let your weight rest on the eight and let the rope slip through your hand and you'll be down in minutes."

"Joan! Hurry up! What's taking so long?" Doug yelled from the top of the tree.

"I'm going!" I yelled back.

"When, next week? I'm freezing and it's getting dark!"

Finally I squatted down and relaxed my weight onto the eight. It held. I began letting the rope slide through my hand.

"Good-bye, Butterfly," I said, looking up at her. "Thank you." Once on the rope, it is easy to feel like an instant pro. I slid for a few feet, braked with a squeeze of my hand and then let myself go, Luna's trunk passing effortlessly before my eyes.

"Good bye, Luna."

When my feet touched the ground, it seemed firmer than I remembered it. On the hike down, I kept pointing to trees in the moonlight, asking Sawyer. "Were we that high?"

"Higher."

"How tall is that tree?"

"About ninety or a hundred feet."

"We were *twice* that high?"

"Yep."

When I got home, I looked up at the trees in my yard and it was as if someone had cast a spell and shrunk them. I looked at the first branch, twenty feet in the air, and smiled.

I said that before I ended this book I would see an old-growth tree cut down. And I did, out in Owl Creek, but I meant one the size of Luna. I felt an obligation to hear that sickening sound of a million-pound tree hitting the earth. It would be my penance. But Doug and I have done something better. We have gone up into an ancient tree's reality. The power of having done this makes my first idea seem simplistic. This is a struggle for life, not death.

CHAPTER 80 TAKE-OUT

LAST night David and I went out to dinner and went bowling. Suzanne was at a friend's for the weekend, so he and I had the luxury of spending some time one-on-one. He wanted to have sushi at a local Japanese restaurant, and I watched him mixing his tamari with his wasabi with such ease that I saw David the young man, rather than David the little boy.

At one point he said, "I forgot what I was going to say."

"Well, it'll come back to you," I reassured.

"Yes, but I was just making conversation before. This was something I really wanted to say." I was amused by his distinctions, and touched that he had been "making conversation" with me, entertaining me, being good company. While we waited for our dinner, he took his disposable wooden chopsticks out of their paper wrapper and was rubbing one on the other to smooth their sides in a sophisticated move he has learned from his sister. I didn't realize he had become so cosmopolitan himself.

"How much do you think these chopsticks cost them? A penny?" he asked.

"Well, more than that," I replied, looking at the lightweight wrapper and the softwood chopsticks.

"I don't think too much more than that," he continued, holding up the wrapper. "This is pretty thin. It doesn't cost much to produce, and neither do the chopsticks."

"Maybe five cents?" I asked.

"I don't think that much," he countered again.

"We should try to find out," I answered, not having the faintest idea how one would research this question.

"If 60 percent of the trees that are cut down go for paper, how do you think that amount is broken down?" he asked me, onto the next thought.

"I don't know . . . " I began.

"Well, let's see, there's computer paper, newspapers, magazines . . .

packaging, toilet paper, paper towels . . . miscellaneous paper . . . " Then he was thinking aloud, throwing around percentages like a wood products broker, while the people at the next table slowed their conversation to listen. He refined his own answers, shifting 10 percent from one category to another, while I thought to myself, "Is this one of the results of this book?" Rather than being a desperately frustrated activist, is he starting at the age of ten to put himself in a position to make a difference in how wood products are consumed?

"And fax paper . . . " I interjected.

"Yeah, well, computer paper . . . " he continued, still figuring.

As we paid our bill, he was pointing behind the counter, whispering, "Miscellaneous paper . . . miscellaneous paper . . . "

As we walked to our car, he asked me, "If take-out boxes were actually made from old-growth redwoods, do you know how many trees this would use a day?"

"Where did you learn this fact?" I stalled, unable to wrap my mind around the question.

"From Ann." This was his math teacher at school. "Six," he went on, answering his own question. I tried to imagine this wood used for something so absurd as take-out boxes, as I pulled my jacket tighter around my body, trying not to crush his leftover sushi in the take-out box that I myself held. While I find the thought of the twenty-first century overwhelming, I felt comforted that for someone like David its problems may be merely enjoyable challenges.

Leather fern, clouded salamander

CHAPTER 81 ENDING

I HAVE reached the day of the deadline for this manuscript. I am going to send it in. The issue of the revocation of PL's license is unresolved. Numerous lawsuits are in progress, but the day has come to stop. I surprised David by suddenly saying, in the middle of what is firmly observed as a work day, "Let's go somewhere."

We have come to the Corkscrew Tree in Redwood National Park. I came here recently as part of a tour for a visiting attorney from Washington, D.C. He wanted to see for himself what the uproar is all about out here in California. Chuck and I flew with him over MAXXAM's land, and then Chuck back to his work while I brought the attorney to see old-growth redwoods in Redwood National Park. Now I sit with David on the same log where the attorney and I ate lunch. David is writing a paper for school, while I am writing about the flight, which was like a reminiscence of the past year.

Before we took off, Chuck showed the lawyer a map of our intended route.

"The dark green is the old-growth," Chuck explained.

"The dark green?" the lawyer clarified.

"Yes," Chuck said matter-of-factly. "We'll be flying—"

"That's it?" the lawyer interrupted. "That's *all* that's left?"

"Yes," Chuck and I said together, spontaneously comforted by the reaction of someone who is not numb to the destruction.

"I just can't believe this. How'd the rest of the map get so . . . white?"

I laughed. The "white" he referred to was the white of the paper within the black line that demarcated MAXXAM's holdings.

"They cut it down," I answered.

"In the past ten years?"

"Yep, most of it," Chuck replied.

Once we were airborne and over clearcuts, we pointed out slash burning, areas recently sprayed with herbicides, the snowy looking forests of pampas grass, and, one by one, the various groves. Meanwhile the lawyer kept saying,

"It's just so . . . gone," over and over. "No one in the rest of the country would believe this. They think, 'Oh well, everything is protected in parks,' or 'It's just a bunch of radical crackpots trying to save every last tree,' or 'The trees'll grow back, so what's the big deal?' But you can't grow back thousand-year-old trees," he informed us, as if we thought we could.

I, meanwhile, became increasingly lost in memory. Instead of seeing the ground below me as simply a faceless wasteland the way I once had, I saw it in terms of the trips Doug and I had taken. We flew toward Allen Creek Grove first, and Chuck asked the pilot to circle around the grove while he gave its statistics. I looked down in silence at the haul road along Yager Creek, where I had felt so vulnerable after David and Doug and I trespassed through Yager Camp. I could even see the stumps of the eighty trees where the eagles used to roost. We passed All Species Grove and there was the log where Chuck had admitted his importance in saving that grove. We reached Shaw Creek Grove, and I studied the road where Doug and I had camped, attempting to fix my eyes on the very place where the canopy shades the salamander pool. We circled over Owl Creek Grove, hit hard, as if in vengeance for the successful lawsuit over the Thanksgiving Massacre. And there was Boot Jack Prairie, where I lay in my sleeping bag at dawn last spring and listened to the varied thrushes sing ownership.

Finally we flew over Headwaters Grove itself, the biggest grove of all. The tops of old-growth trees spired toward us and momentarily filled the frame of my window, allowing my gaze to rest on their even beauty. To my surprise, I found myself saying silently, "I want to go back there." I couldn't believe it. I have finally seen all the groves, yet apparently I am not ready to stop. I laughed as I imagined telling Doug that I want to go back to Headwaters. My conversion must be complete.

But as we kept flying, the grove abruptly ended, and I found myself staring down once more at slope after slope of clearcuts, registering solemn dismay that even though I am done with my writing, the clearcuts are not like characters in fiction. They do not remain frozen when the book covers are closed. I have turned off my computer, but the saws keep running.

EPILOGUE

WHEN I was in my twenties I had a friend named Chuy with whom I was working to try to prevent the extension of the Santa Barbara airport into the marsh that surrounded the University of California where he and I were attending college. At one point, when all of our arguments and pleas seemed to be falling on deaf ears, I turned to Chuy and I said,

"What's the use? We're doomed anyway."

Chuy dropped his normal, crooked grin, looked at me very seriously, and said, "I'll go down fighting."

Anyone could have said this to me, but Chuy said it first, and it became embedded in my personal philosophy. Inactivity due to dispair is not an option. Whenever I start to ask myself, "What's the use," I counter with, "I'll go down fighting."

However one says it—"Action is the antidote to dispair." "Find a corner of the world and fix it." "If the people lead, the leaders will follow."—the important thing is that we not live in denial, that we investigate, communicate, and act meaningfully, in proportion to the crisis at hand—that is, the destruction of our planet.

What can you do?

Keep an open mind. Use your imagination. If anyone would have told me two years ago that I would have climbed a 180-foot redwood tree to conduct an interview, I wouldn't have believed it. Brainstorm with other people. Use your work as a way to create meaningful relationships that will sustain you and reward you.

In a little more than five hundred years we have wiped out over 90 percent of North America's ancient forests starting from the east coast and moving west (see map, p. viii). Only 3.5 percent of our ancient redwood forests now remain.

The Clinton administration's Deal would protect only about one tenth of the 60,000-acre Headwaters Forest. Only two out of the six major groves would be permanently protected, while the other four are temporarily protected under the Endangered Species Act. That temporary protection is only as strong as the Endangered Species Act itself, which is currently up for reauthorization. The remaining acreage of Pacific Lumber's land would continue to be logged, slash-burned, and herbicide-sprayed. Meanwhile, much of the acreage that has been severely logged needs massive restoration to control erosion and remove non-native plants and promote the growth of new forest.

As this book goes to press . . .

Butterfly is still up in Luna—six months and counting.

The residents of Stafford, Freshwater, and Elk River have filed suit against PL.

The victims of the pepper spraying incidents have filed suit against Humboldt County and the City of Eureka based on the actions of the sheriff's department and the Eureka police department. The sheriff's department continues to issue cotton balls and pepper spray to its officers for use against protestors.

The Office Thrift Supervision has filed suit against Charles Hurwitz to recover the money the United States government has spent to bail out the failed United Savings Association of Texas. The trial is being held in Houston, Texas.

The future of the "Deal" remains unresolved. Of $380 million that is to be paid to Charles Hurwitz, the federal share, which is $250 million, has been appropriated. But the appropriation of California's share of $130 million is currently delayed in the state senate because many senators believe that the protections proposed for the threatened and endangered species within Headwaters are inadequate. Without the state's share of the money, the "Deal" may collapse.

Pacific Lumber has released their draft Habitat Conservation Plan for public comment. If all goes as planned, PL will be granted permission to log some of their old-growth stands occupied by nesting marbled murrelets as early as the fall of 1999. The HCP may protect for only fifty years the main groves in Headwaters with the exception of Owl Creek Grove. Currently

Owl Creek Grove and the murrelet groves around Grizzly Creek State Park continue to be the objects of horse-trading between the government and Hurwitz. Any of the rest of PL's 214,000 acres that have any trees still standing remain vulnerable to being clear-cut, slash-burned, and sprayed with herbicides.

It appears likely that environmental groups may litigate the HCP if no additional meaningful measures are added for the protection of the species at risk.

A lot of tired people need your support.

RESOURCES AND REFERENCES

Here are some suggestions for actions you can pursue:

Write, call, or fax President Clinton, Senator Dianne Feinstein, and your federal representatives in Congress to voice your support of public acquisition of the entire 60,000-acre Headwaters Forest.

President Bill Clinton
The White House
1600 Pennsylvania Avenue NW
Washington, DC 20500
president@whitehouse.gov

Senator Dianne Feinstein
Senate Office Building
Washington, DC 20510
(202) 224-3841
(415) 249-4777
fax: (202) 228-3954
senator@feinstein.senate.gov

Your senator—
Senate Office Building
Washington, DC 20510
senator@[your senator's last name].senate.gov

- Urge the President and federal representatives in Congress to reauthorize and strengthen the Endangered Species Act, which will increasingly come under attack as natural resources become even more scarce.
- Work for a ban on the logging of all ancient forests on private and public lands, including complete protection for all roadless areas on our public lands. Most of the remaining old growth is on public lands in our National Forests. We are losing tens of millions of tax dollars every year by subsidizing the timber industry while it levels our forests.
- Work to ban clear-cutting in California and forests nationwide.
- Buy wood products only where *certified sustainably harvested* lumber is sold.
- Work to change laws to create wider stream zones, and insist on enforcement of stream-protection laws.
- Investigate local and national environmental issues before voting in elections. Help educate your friends about the issues before they cast their ballots.

Here are some names and addresses that will help you stay in touch and get involved:

Doug Thron
Thron Nature Photography
PO Box 703
Arcata, CA 95518
(707) 822-4870
To schedule Headwaters slideshow presentations and to receive a monthly newsletter

Bay Area Action
715 Colorado Avenue, Suite 1
Palo Alto, CA 94303
(650) 321-1994
www.baaction.org

Bay Area Coalition for Headwaters (BACH)
2530 San Pablo Avenue
Berkeley, CA 94702
(510) 548-3113
bach@igc.org; www.igc.apc.org/ headwaters/bach.html

Californians for Alternatives to Toxics
PO Box 1195
Arcata, CA 95518
(707) 822-8497; fax: (707) 822-7136
cats@igc.org; http:/www.mapcruzin.com/cats

Democracy Unlimited
PO Box 27
Arcata, CA 95518
(707) 822-2242; fax: (707) 822-3481
duhc@monitor.net; www.monitor.net/duhc
Write to receive booklist, newsletter about workshops, introductory packet on democracy and corporations

Earth First! (Redwood Nation)
106 W. Standley Street
Ukiah, CA 95482
(707) 468-1660
www.net-code.com/headwaters/

Earth Island Institute
300 Broadway, Suite 28
San Francisco, CA 94133
(415) 788-3666
Environmental organizing and publishing network

Environmental Protection Information Center (EPIC)
PO Box 397
Garberville, CA 95542
(707)923-2931
epic@igc.org; www.igc.org/epic/
To receive their newsletter, "Wild California" and other information on the protection of Headwaters

Headwaters Action Video Collective
PO Box 2198
Redway, CA 95560
(707) 459-5490
To order a compelling video of Butterfly's tree-sit in Luna

Headwaters Forest Website
www.headwatersforest.org/

Humboldt Watershed Council
828 G Street
Eureka, CA 95501
(707) 443-7433; fax: (707) 443-3331
sheds@humboldt1.com
Monthly newletter about the Headwaters issue; excellent video, "Voices of Humboldt County"

Institute for Sustainable Forestry (ISF)
PO Box 1580
Redway, CA 95560
(707) 247-1101; fax: (707) 247-3555
info@isf-sw.org
To receive their newsletter, "Forestree News" and for information on forest management, watershed restoration, Wild Iris Forest Products, SmartWood Certification Program

Jail Hurwitz Website
www.jailhurwitz.com/

League of Conservation Voters
1707 L Street NW, Suite 750
Washington, DC 20036
(202) 785-8683
To obtain the environmental voting records for legislators

Luna Media Services
www.humboldt1.com/~lunanews

Native Forest Council
PO Box 2190
Eugene, OR 97402
(541) 688-2600
Works for the protection of National Forests

Northcoast Environmental Center
879 Ninth Street
Arcata, CA 95521
phone/fax: (707) 822-0827
nec@northcoast.com
To receive environmental newsletter "The ECONEWS"

Program on Corporations, Law and Democracy
211.5 Bradford Street
Provincetown, MA 02657
phone/fax: (508) 487-3151

Project Lighthawk
P.O. Box 29231; Presidio Building, 1st floor
San Francisco, CA 94129-0231
(415) 561-6250
Assistance to groups needing access to airplanes and pilots for aerial surveys, photography, and press coverage of endangered environmental areas

Rainforest Action Network
221 Pine Street #500
San Francisco, CA 94104
(415) 398-4404

Rose Foundation for Community and the Environment
6008 College Avenue, Suite 10
Oakland CA 94618
(510) 658-0702; fax: (510) 658-0732
This organization focuses on "debt for nature" as a means to preserve Headwaters Forest

Save America's Forest
4 Library Court SE
Washington, DC 20003
(202) 544-9219
Works for the protection of National Forests

Sierra Club
85 2nd Street, 2nd floor
San Francisco, CA 94105-3459
(415) 977-5500

Siskyou Project
PO Box 220
Cave Junction, OR 97523
(541) 592-4459; fax: (541) 592-2653
Works for the protection of National Forests

Taxpayers for Headwaters Forest
PO Box 456
Bayside, CA 95524
(707) 825-0444
www.laststand.org/

Trees Foundation
PO Box 2202
Redway, CA 95560
(707) 923-4377; fax: (707) 923-4427
trees@igc.org
Assistance in grassroots environmental activism, including design, writing, and production of educational outreach materials; training in organizational skills from fundraising to budget development

Western Ancient Forest Campaign
1025 Vermont Avenue NW, 3rd floor
Washington, DC 20005
(202) 879-3188; fax: (202) 879-3189
Works for the protection of National Forests

Here is a selection of books pertaining to Headwaters and the ancient forests:

Bari, Judi. *Timber Wars*. Monroe, Maine: Common Courage Press, 1994.

Brown, Joseph E. *Monarchs of the Mist: The Story of Redwood Natural Park and the Coast Redwoods*. Point Reyes, California: Coastal Parks Association, 1982.

Byans, Gerald. "Stories of a Logger's Life," *The Humboldt Historian* (Humboldt County Historical Society) 45, no. 5 (Fall 1997): 29.

Camp, Orville. *The Forest Farmer's Handbook: A Guide to Natural Selection Forest Management*. Ashland, Oregon: Sky River Press, 1984.

Chaney, Ralph. *Redwoods of the Past*. San Francisco: Save-the-Redwoods League, c. 1930.

Corkran, Charlotte C., and Chris Thomas. *Amphibians of Oregon, Washington, and British Columbia*. Vancouver: Lone Pine Publishing, 1998.

Crescoe, Audrey. *Giants: The Colossal Trees of Pacific North America*. Boulder, Colorado: Roberts Rinehart Publishers, 1997.

Davis, Wade. *Rainforest: Ancient Realm of the Pacific Northwest*. With photographs by Graham Osborne. White River Junction, Vermont: Chelsea Green, 1999.

Devall, Bill. *Clearcut: The Tragedy of Industrial Forestry*. San Francisco: Sierra Club Books/Earth Island Press, 1993.

Efert, Larry, and Bob Normann. *The Tall Trees*. Fortuna, California: FVN Corporation, 1997.

Fritz, Emanuel. *Story Told by a Fallen Giant*. San Francisco: Save-the-Redwoods League, c. 1930.

Hams, David. *The Last Stand: The War Between Wall Street and Main Street over California's Ancient Redwoods*. New York: Times Books, 1996.

Hewes, Jeremy Joan. *Redwoods: The World's Largest Trees*. New York: Smithmark Publishers, 1981.

Jepson, Willis Linn. *Trees, Shrubs and Flowers of the Redwood Region*. San Francisco: Save-the-Redwoods League, 1954.

Keator, Glenn, and Ruth M. Heady. *Pacific Coast Fern Finder*. Berkeley: Nature Study Guide, 1981.

Kelly, David, and Gary Braasch. *Secrets of the Old Growth Forest*. Salt Lake City: Peregrine Smith Books, 1988.

Kozloff, Eugene N. *Plants and Animals of the Pacific Northwest: An Illustrated Guide to the Natural History of Western Oregon, Washington, and British Columbia*. Seattle: University of Washington Press, 1978.

Leaders Magazine, Inc. "Value Visionary: An Interview with Charles B. Hurwitz, Chairman, President and Chief Executive Officer, MAXXAM Inc. and MAXXAM Group Inc., Houston, Texas," *Leaders* 17, no. 2 (April, May, June 1994): 48.

Leonard, William F. *Amphibians of Washington and Oregon*. Seattle: Seattle Audubon Society, 1995.

Lewis, Adam. *Salmon of the Pacific*. Seattle: Sasquatch Books, 1994.

Leydet, François. *The Last Redwoods and the Parkland of Redwood Creek*. San Francisco: Sierra Club Books, 1989.

Lyons, Kathleen, and Mary Beth Cuneo-Lazaneo. *Plants of the Coast Redwood Region*. Los Altos, California: Looking Press, 1988.

Maser, Chris. *Forest Primeval: The Natural History of an Ancient Forest*. San Francisco: Sierra Club Books, 1989.

Niehaus, Theodore F., and Charles L. Ripper. *Pacific States Wildflowers*. Boston: Houghton Mifflin, 1976.

Pojar, Jim, and Andy MacKinnon. *Plants of the Pacific Northwest Coast: Washington, Oregon, British Columbia and Alaska*. Vancouver: Lone Pine Publishing, 1994.

Rohde, Jerry and Cisela Rohde. *Humboldt Redwoods State Park: The Complete Guide*. McKinleyville, California: MountainHome Books, 1992.

Stebbins, Robert C., and Nathan W. Cohen. *A Natural History of Amphibians*. Princeton: Princeton University Press, 1995.

Whitaker, John O. *National Audubon Society Field Guide to North American Mammals*. New York: Alfred A. Knopf, 1996.

Whitney, Stephen. *Western Forests*. New York: Alfred A. Knopf, 1986.